SWITCHED-CURRENT
SIGNAL PROCESSING AND
A/D CONVERSION CIRCUITS

THE KLUWER INTERNATIONAL SERIES IN ENGINEERING AND COMPUTER SCIENCE

ANALOG CIRCUITS AND SIGNAL PROCESSING
Consulting Editor: **Mohammed Ismail**. *Ohio State University*

Related Titles:

RESEARCH PERSPECTIVES ON DYNAMIC TRANSLINEAR AND LOG-DOMAIN CIRCUITS
    W.A. Serdijn, J. Mulder
    ISBN: 0-7923-7811-3
CMOS DATA CONVERTERS FOR COMMUNICATIONS
    M. Gustavsson, J. Wikner, N. Tan
    ISBN: 0-7923-7780-X
DESIGN AND ANALYSIS OF INTEGRATOR-BASED LOG -DOMAIN FILTER CIRCUITS
    G.W. Roberts, V. W. Leung
    ISBN: 0-7923-8699-X
VISION CHIP
    A. Moini
    ISBN: 0-7923-8664-7
COMPACT LOW-VOLTAGE AND HIGH-SPEED CMOS, BiCMOS AND BIPOLAR OPERATIONAL AMPLIFIERS
    K-J. de Langen, J. Huijsing
    ISBN: 0-7923-8623-X
CONTINUOUS-TIME DELTA-SIGMA MODULATORS FOR HIGH-SPEED A/D CONVERTERS: Theory, Practice and Fundamental Performance Limits
    J.A. Cherry, W. M. Snelgrove
    ISBN: 0-7923-8625-6
LEARNING ON SILICON: Adaptive VLSI Neural Systems
    G. Cauwenberghs, M.A. Bayoumi
    ISBN: 0=7923-8555-1
ANALOG LAYOUT GENERATION FOR PERFORMANCE AND MANUFACTURABILITY
    K. Larnpaert, G. Gielen, W. Sansen
    ISBN: 0-7923-8479-2
CMOS CURRENT AMPLIFIERS
    G. Palmisano, G. Palumbo, S. Pennisi
    ISBN: 0-7923-8469-5
HIGHLY LINEAR INTEGRATED WIDEBAND AMPLIFIERS: Design and Analysis Techniques for Frequencies from Audio to RF
    H. Sjöland
    ISBN: 0-7923-8407-5
DESIGN OF LOW-VOLTAGE LOW-POWER CMOS DELTA-SIGMA A/D CONVERTERS
    V. Peluso, M. Steyaert. W. Sansen
    ISBN: 0-7923-8417-2
THE DESIGN OF LOW-VOLTAGE, LOW-POWER SIGMA-DELTA MODULATORS
    S. Rabii, B.A. Wooley
    ISBN: 0-7923-8361-3
TOP-DOWN DESIGN OF HIGH-PERFORMANCE SIGMA-DELTA MODULATORS
    F. Medeiro, A. Pérez-Verdú, A. Rodríguez-Vázquez
    ISBN: 0-7923-8352-4
DYNAMIC TRANSLINEAR AND LOG-DOMAIN CIRCUITS: Analysis and Synthesis
    J. Mulder, W. A. Serdijn, A.C. van der Woerd, A.H.M. van Roermund
    ISBN: 0-7923-8355-9

SWITCHED-CURRENT SIGNAL PROCESSING AND A/D CONVERSION CIRCUITS

Design and Implementation

by

Bengt E. Jonsson
Ericsson Radio Systems AB

KLUWER ACADEMIC PUBLISHERS
BOSTON / DORDRECHT / LONDON

A C.I.P. Catalogue record for this book is available from the Library of Congress.

ISBN 0-7923-7871-7

Published by Kluwer Academic Publishers,
P.O. Box 17, 3300 AA Dordrecht, The Netherlands.

Sold and distributed in North, Central and South America
by Kluwer Academic Publishers,
101 Philip Drive, Norwell, MA 02061, U.S.A.

In all other countries, sold and distributed
by Kluwer Academic Publishers,
P.O. Box 322, 3300 AH Dordrecht, The Netherlands.

Printed on acid-free paper

All Rights Reserved
© 2000 Kluwer Academic Publishers, Boston
No part of the material protected by this copyright notice may be reproduced or
utilized in any form or by any means, electronic or mechanical,
including photocopying, recording or by any information storage and
retrieval system, without written permission from the copyright owner.

Printed in the Netherlands.

*To Cristina, my wife and best friend
for endless support and lots of fun*

Contents

About the author ... xi

Acknowledgements ... xiii

Preface ... xv

1. Introduction ... 1
   1.1 TECHNICAL CHALLENGES IN MIXED-SIGNAL INTEGRATION ... 1
   1.2 THE SWITCHED-CURRENT TECHNIQUE ... 2
   1.3 ABOUT THIS BOOK ... 4
      Part I: An introduction to the SI technique ... 4
      Part II: SI circuit design issues ... 5
      Part III: SI circuit implementation examples ... 5
      Appendix ... 6
   REFERENCES ... 6

Part I: An Introduction to the SI Technique ... 7

2. Switched-Current Circuits ... 9
   2.1 CURRENT-MODE BUILDING BLOCKS ... 9
      2.1.1 The current mirror ... 9
      2.1.2 First and second generation SI memory cells ... 10
      2.1.3 OTA based current S/H circuits ... 12
      2.1.4 Comparators ... 12
      2.1.5 D/A-converters ... 13
   2.2 NON-IDEAL EFFECTS ... 14
      2.2.1 MOS transistor mismatch ... 15

		2.2.2	Conductance ratio errors	18
		2.2.3	Clock-feedthrough	19
		2.2.4	Noise	24
		2.2.5	Settling	30
		2.2.6	Voltage drop	31
		2.2.7	Jitter	31
		2.2.8	Summary of non-ideal effects	32
	2.3	CLOCK-FEEDTHROUGH COMPENSATION		32
		2.3.1	Attenuation techniques	32
		2.3.2	Cancellation techniques	33
		2.3.3	Algorithmic techniques	36
		2.3.4	Fully-differential, feedforward and feedback techniques	37
		2.3.5	Adaptive techniques	38
		2.3.6	Zero-voltage switching techniques	38
		2.3.7	CFT compensation techniques used in this book	39
	2.4	CURRENT SAMPLE-AND-HOLD EVOLUTION		39
	REFERENCES			42
3.	Switched-Current Systems			49
	3.1	SAMPLED-DATA FILTERS		49
		3.1.1	FIR-filters	49
		3.1.2	IIR-filters	50
		3.1.3	Miscellaneous filters	53
		3.1.4	Design methodology and analysis	53
		3.1.5	Influence of SI circuit imperfections	54
	3.2	A/D CONVERTERS		54
		3.2.1	Nyquist A/D converters	55
		3.2.2	Oversampling Δ-Σ A/D converters	59
		3.2.3	Influence of SI circuit imperfections	60
	REFERENCES			61

Part II: SI Circuit Design Issues   69

4.	Clock-Feedthrough Compensated First-Generation SI Circuit Design			71
	4.1	INTRODUCTION		71
	4.2	CLOCK-FEEDTHROUGH COMPENSATION IN SI CIRCUITS		72
		4.2.1	Clock-feedthrough modeling	72
		4.2.2	Large-signal CFT error current	73
		4.2.3	Complete CFT cancellation	74
		4.2.4	Coefficient matching sensitivity	76
	4.3	MEMORY CELL DESIGN OPTIONS		77
		4.3.1	Optimizing an arbitrary goal function	78
		4.3.2	Minimizing CFT	79

	4.4 EXPERIMENTAL RESULTS	80
	4.5 SUMMARY	80
	REFERENCES	81
5.	Sampling Time Uncertainty	83
	5.1 INTRODUCTION	83
	5.2 SAMPLING TIME UNCERTAINTY	84
	5.3 REDUCING SAMPLING TIME UNCERTAINTY	86
	5.3.1 Random jitter	86
	5.3.2 Signal dependent jitter	87
	5.4 SUMMARY	89
	REFERENCES	90
6.	Design of Power Supply Wires	91
	6.1 INTRODUCTION	91
	6.2 VOLTAGE DROP ON POWER SUPPLY WIRES	92
	6.3 DESIGN CONSIDERATIONS	94
	6.3.1 Quiescent point shift	94
	6.3.2 Offset and distortion	95
	6.4 SUMMARY	96
7.	SI Circuit Layout	97
	7.1 INTRODUCTION	97
	7.2 DIFFERENCE BETWEEN SI AND SC CIRCUIT LAYOUT	97
	7.3 FLOORPLANNING	99
	7.4 LAYOUT STYLES	100
	7.4.1 Style I	100
	7.4.2 Style II	102
	7.5 LAYOUT AUTOMATION	103
	REFERENCES	105
Part III: SI Circuit Implementation Examples		107
8.	A 3.3-V CMOS Wave SI Filter	109
	8.1 INTRODUCTION	109
	8.2 DISCRETE-TIME WAVE FILTERS	110
	8.3 CURRENT-MODE REALIZATION OF N-PORT ADAPTORS	112
	8.4 SWITCHED-CURRENT DELAY ELEMENT REALIZATION	113
	8.5 POTENTIAL FOR AUTOMATIC GENERATION	115
	8.6 FILTER REALIZATION	117
	8.7 SIMULATION RESULTS	118
	8.8 EXPERIMENTAL RESULTS	119
	8.9 SUMMARY	119
	REFERENCES	120

9. **A 3.3-V CMOS Switched-Current Delta-Sigma Modulator** — 121
   - 9.1 INTRODUCTION — 121
   - 9.2 MODULATOR STRUCTURE — 122
   - 9.3 SYSTEM LEVEL SIMULATIONS — 123
   - 9.4 CIRCUIT IMPLEMENTATION — 124
   - 9.5 EXPERIMENTAL RESULTS — 125
   - 9.6 SUMMARY — 127
   - REFERENCES — 127

10. **A 3-V Wideband CMOS Switched-Current A/D-Converter** — 129
    - 10.1 INTRODUCTION — 129
    - 10.2 SWITCHED-CURRENT A/D CONVERTERS — 131
    - 10.3 RSD A/D CONVERTER ARCHITECTURE — 132
    - 10.4 CIRCUIT IMPLEMENTATION — 133
    - 10.5 SIMULATION RESULTS — 136
      - 10.5.1 System level simulations — 136
      - 10.5.2 Circuit level simulations — 138
    - 10.6 EXPERIMENTAL RESULTS — 140
    - 10.7 PERFORMANCE COMPARISON — 144
    - 10.8 SUMMARY — 147
    - REFERENCES — 148

11. **A Dual 3-V 32-MS/s CMOS Switched-Current ADC** — 151
    - 11.1 INTRODUCTION — 151
    - 11.2 A/D CONVERTER ARCHITECTURE — 152
    - 11.3 CIRCUIT IMPLEMENTATION — 153
    - 11.4 EXPERIMENTAL RESULTS — 154
      - 11.4.1 ADC core cell — 154
      - 11.4.2 Parallel ADC — 156
    - 11.5 SUMMARY — 158
    - REFERENCES — 159

12. **Conclusions** — 161

Appendix: Noise Integrals — 165

Index — 167

About the author

Bengt Jonsson received the M.Sc. degree in Electrical Engineering from *Linköping University*, Sweden. He also received the Licentiate degree in Applied Electronics 1994 from the same university. In 1996 he joined *Ericsson Radio Systems AB*, and in 1999 he received the Ph. D. degree in Electronic System Design from the *Royal Institute of Technology*, Stockholm, Sweden. He is currently with *Ericsson Radio Systems AB*, working as a Senior Specialist in Data-Conversion Technology. His main research interests are A/D and D/A converter design, including switched-capacitor and switched-current circuits for high-speed, high-resolution data converters.

Acknowledgements

During my time as a Ph.D. student I had three supervisors, and each one of them shared of his knowledge and gave me important insight in different areas. Professor Lars Wanhammar of Linköping University has educated me in ASIC design and digital signal processing. Although these topics are not covered in the book, the knowledge has been useful for the work included in the book, and for my present work in industry. The late Dr. Sven Eriksson, also of Linköping University, introduced me to the field of analog CMOS design, and the switched-current technique. It was very inspiring to get the opportunity to work with a new circuit technique as it was emerging. Finally, I'm grateful that Professor Hannu Tenhunen at the Royal Institute of Technology, Stockholm, was willing to help me finish under his supervision. His comments and suggestions have been most helpful and encouraging during the second half of the Ph.D. work that this book is based upon.

It has been a great pleasure working for many years with Dr. Nianxiong Tan, who is now with GlobeSpan Inc., NJ. His friendship, his professional skills, and his support in good and bad times is deeply appreciated. He is also the co-author of chapter 9. During my carrier in electronics design, there has always been an abundance of good colleagues around. In Linköping, I have enjoyed working close to Håkan Träff, Dr. Mikael Gustavsson, Jacob Wikner, and the rest of the staff at the Applied Electronics group. I wish to thank Dr. Mark Vesterbacka for being an excellent brainstorming partner, Dr. Kent Palmkvist for always helping me out with Macintosh trouble, and Mats Larsson for being a fantastic UNIX administrator and a supportive friend. Dr. Christer Jansson, Dr. Jan-Erik Eklund, and the entire Electronic Devices group, led by Professor Christer Svensson, have also contributed with important knowledge in CMOS implementation.

A very pleasant working environment is created by my present colleagues at Ericsson Radio Systems, Giti Amozandeh, Helge Stenström, Per Lundborg, Mikael Pettersson, and the rest of the fantastic *RF IC & Mixed-Signal Design* group managed by John Lundquist. I'm grateful to my previous boss, Tomas Melander, who gave me time and resources to spend on the chip measurements reported in chapters 10 and 11. During these measurements, I also got valuable assistance from Dr. Mats Johansson from the Radio Research group, and from my colleagues Helge Stenström and Jerry Lundholm.

I want to express my gratitude to my parents Karin and Hilding for all their love and support. I'm glad to say that they let me grow up and develop in a positive atmosphere that ensured me that I would always be loved for who I am – with or without academic degrees. I nearly gave up my Ph.D. work as I met my wife Cristina, realizing that there are far more important things in life than degrees. However, thanks to her admirable patience and endless support, I've managed to find the time and inspiration to finish both the Ph.D. work **and** this book. *A good wife is truly a gift from God*! I also appreciate the wealth of friends and relatives in Sweden and Portugal for the various ways they have enriched my life.

– "It is God who arms me with strength and makes my way perfect"
2 Sam. 22:33

Preface

This book covers several aspects of the design and implementation of switched-current (SI) circuits for signal filtering and A/D conversion. Most of the material in the book originates from the author's Ph.D. thesis. The text has been reorganized to better fit the format of a book. Recent literature references and some additional work have also been added. Reading this book requires no prior knowledge of SI circuits. It is only assumed that the reader is familiar with basic analog circuit design. The first three chapters introduce the reader to the switched-current technique, providing a rich set of references for further reading. These references have been carefully selected from the author's near-exhaustive collection of papers on the topic. It has been the author's aim that readers at any level of experience should find the reference collections to be a valuable resource – something that the reader would often return to.

One may choose to read this book for a variety of reasons: The text is suitable for analog designers wishing to familiarize themselves with the SI technique without getting too serious. Due to the simplicity of most SI circuits, it should be easy enough to get a feeling for the basic SI circuit functionality, their most critical design issues, and typical applications. An intermediate level reader may be a professional designer, or a Ph.D. student, starting to design switched-current circuits. Such a reader will find a lot of information that can be used as a starting point for his or her own work. Without being a cookbook, the text includes useful design formulas where appropriate, and the design examples in chapters 8 through 11 should give the reader some direction. Finally, the experienced SI circuit professional is most likely familiar with the majority of topics presented in this book. Nevertheless, he or she should be able to find enough performance and design details in the implementation examples to find the text worth reading. It should also be noted that chapters 5 through 7 deal with design issues not usually covered in SI publications: sampling time uncertainty, power supply voltage drops, and layout automation. For the switched-current A/D converter designer there are a few performance comparison charts in chapter 10.

Most of the principles and design examples outlined in this book can, and should, be applied to second-generation as well as first-generation SI circuits. Yet the author admits that this book is clearly biased toward first-generation SI circuits in its derivations and examples. This is due to the fact that the author has mainly been using this type of circuit throughout his work. For many years, most of the development in SI circuits has been for second-generation circuits, and today they are probably a better choice. In spite of the advantages of second-generation circuits, such as reduced power dissipation and mismatch problems, the performance demonstrated by some of the circuits in this book show the usefulness of first-generation SI circuits.

Bengt E. Jonsson

Stockholm, March 2000

Chapter 1

Introduction

Information technology (IT) has demonstrated a tremendous growth in market value as well as in technical performance. The technical catalyst in the IT explosion is the constantly improved digital VLSI technology, where packing density, low-power and low cost have been key issues. As a result of this development, it has become possible to integrate extremely complex digital systems onto a single chip. The next goal is to enable *one-chip system solutions* by successfully integrate *interfacing circuits* such as A/D and D/A converters on the same chip as well. Developing analog interfacing circuits for *mixed analog-digital integration* is of fundamental importance to the future advance of microelectronics and information technology. First of all, the world around us – our ultimate source of information – is essentially *analog* in nature. Whenever an electronic system is interacting with the real world there is a need for interfacing circuits. Figure 1.1 illustrates some of the interaction between the analog and digital realms in today's technology. Secondly, in order to cut manufacturing cost, which is particularly important for consumer products, and to minimize the power dissipation and weight of portable systems, the ultimate goal for many products is to have one-chip solutions whenever possible.

1.1 Technical challenges in mixed-signal integration

Unfortunately, when manufacturing technology is optimized for digital VLSI, the result is that analog circuit design becomes even more difficult. In addition to problems that has always faced analog designers, the supply voltage is now reduced, which increases the problem with thermal noise and the use of sampling switches. Furthermore, precision capacitors and resistors are usually excluded from the pure digital CMOS process because of the extra manufacturing cost. Analog interfacing circuits that are to be integrated with state-of-the-art digital circuits should therefore preferably not require anything but MOS transistors, and their dynamic range should not be limited by the supply voltage. The *current-mode* approach [1], and the *switched-current* technique [2], were proposed in order to meet these demands.

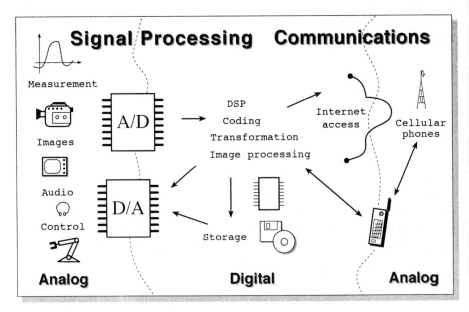

Fig. 1.1 The analog-digital-analog world of today.

1.2 The switched-current technique

The *switched-current* (*SI*) technique for analog signal processing gained interest because of its ability to adapt to the digital VLSI trends, i.e., fine line CMOS processes with low power supply voltages and no extra mask steps for precision analog devices [1-2]. In *current-mode* circuits, signals are represented by currents instead of voltages. Therefore the *signal swing* is only *indirectly* limited by a reduction of the available voltage range. In a traditional *voltage-mode* circuit, the supply voltage imposes a *direct* limitation on signal swing. Current-mode circuits could therefore be a better choice as power supply voltages are lowered. It should however be noted that, since the sampling actually takes place in the voltage domain, the *dynamic range* is limited by kT/C-noise also in SI circuits. The dynamic range of an SI circuit is therefore limited by the power supply voltage just as much as its *switched-capacitor* (SC) counterparts. Current-mode circuits are also claimed to have an inherent potential for *high speed* operation. In SI circuits, there is no fundamental need for high-gain stages, as in SC circuits. The low-gain current amplifiers can be designed so that there are exclusively *low-impedance nodes* in the signal path, thus pushing the dominant poles to much higher frequencies. As will be demonstrated in the remainder of this book, current-mode data-conversion and signal processing circuits suitable for low-voltage operation, can be realized using MOS transistors only. It is also shown that such circuits have very regular structures, and are more suitable for automated design than many voltage-mode analog circuits. This regularity, and the MOS-only nature of SI circuits indicate that they could be used as building blocks for one-chip mixed-signal systems. Switched-current circuits were proposed as late as 1988 [3], and there are still a number of technical problems to be solved. The most serious problem is undoubtedly *charge injection* or *clock-feedthrough* (*CFT*). A number of solutions have been proposed over the years, including one by the author [4]. Several other methods have been proposed – most of them listed in chapter 2. Additional problems including noise, conductance ratio errors and mismatch, are also described in chapter 2.

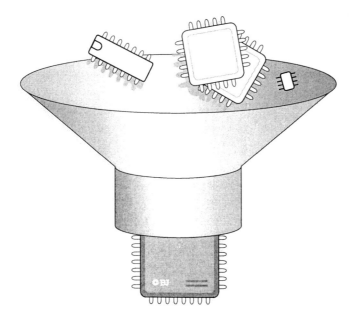

Fig. 1.2 Mixed-signal one-chip solutions.

1.3 About this book

The topic of this book is the design and implementation of switched-current signal processing and A/D conversion circuits. The text is divided into three parts. An *introduction to the SI technique* is given in the **first part**, and the technical problems associated with switched-current circuits are described. A thorough review of the evolution of SI circuits in general, as well as SI filters and A/D converters, is included in chapters 2 and 3. A significant amount of work has been put into selecting the large number of literature references included. It is the author's hope that this reference collection will be a valuable resource to anyone studying the switched-current technique. The **second part**, deals with *a selection of SI circuit design issues* such as clock-feedthrough, sampling time uncertainty, layout design, and design automation. The selection of topics mostly reflects the author's own work regarding SI circuit design and theory, and is far from exhaustive. Several other important design issues are mentioned in chapters 2 and 3, where the reader is given suggestions for further reading. Four *design and implementation examples* are given in the **third part** of the book: A wave SI filter, a $\Delta\text{-}\Sigma$ modulator, a wideband A/D converter core, and a high-speed parallel ADC, all designed and implemented by the author[1]. While chapters 8 – 11 are not specifically written as design tutorials, there is still enough detail for a reader having basic analog design skills to repeat each design. This could be a suitable "getting-started" exercise for the serious reader. Below is an outline of the rest of the book.

Part I: An introduction to the SI technique

Chapter 2 presents *fundamental SI circuits* with respect to their ideal operation. A résumé of non-ideal effects in switched-current circuits is included, as well as a detailed description of switch-induced errors, also known as clock-feedthrough (CFT). Clock-feedthrough models and commonly used clock-feedthrough reduction techniques are reviewed. Finally, the evolution of the SI memory cell as seen in the open literature is summarized on a year-by-year basis.

Chapter 3 gives an overview of the evolution of *SI systems* such as sampled-data filters, A/D converters and oversampling A/D converters, as reported in the open literature. It gives the context for the implementation examples described in chapters 8 – 11.

---

[1] The $\Delta\text{-}\Sigma$ modulator was designed and implemented in cooperation with Dr. Nianxiong Tan, author of *Switched-Current Design and Implementation of Oversampling A/D Converters*, Kluwer, 1997.

Introduction 5

Part II: SI circuit design issues

Chapter 4 describes clock-feedthrough compensation and the design of first-generation SI circuits using a significantly improved memory cell developed by the author. The described circuit was the first SI memory cell that could completely remove both constant and signal-dependent CFT errors. Its clock-feedthrough cancellation scheme is treated in detail, and it is shown how the memory cell can be optimized for power, speed, or accuracy. The performance is compared with the performance of two previously published memory cells using HSPICE™ simulations. Measured results are also included.

Chapter 5 is focusing on the distortion in sampling caused by the signal-dependent switch-off time of MOS switches. It is shown that such distortion can be suppressed in fully-differential sampling circuits on condition that complementary (CMOS) sampling switches are **not** used. These results are applied to SI memory cells, and the theoretical results are verified by MATLAB™ simulations.

Chapter 6 deals with on-chip power-supply interconnections. It is shown that a resistive voltage drop on VDD and ground wires can easily corrupt the performance of a current-mode circuit. Design formulas are derived and MATLAB™ simulations are included. The design examples that are given are related to the design of the ADC core in chapter 10 and the parallel ADC in chapters 11.

Chapter 7 illustrates a few aspects of SI circuit layout design that differs from the layout of switched-capacitor circuits: placement of the sampling capacitor and switches, the regular structure of the SI circuits, and the increased potential for design-automation. The use of CAD tools to shorten design time and reduce the cost for analog ICs is given a brief discussion. A design style, and the structure of a CAD tool suitable for automatic layout generation of SI circuits is suggested. A simplified version of the tool was used for the circuit in chapter 8, and for parts of the circuit in chapter 9.

Part III: SI circuit implementation examples

Chapter 8 describes the implementation of a wave SI filter using general N-port adaptor circuits developed by the author as building blocks. The suitability for automated layout generation of SI circuits is pointed out. Due to process failure, the chip was not functional and could not be measured.

Chapter 9 reports a $\Delta\text{-}\Sigma$ modulator implementation using the clock-feedthrough compensated memory cells described in chapter 4. Measurement results are presented, showing an 11-bit dynamic range and

60 dB peak SNDR. The performance was calculated assuming an OSR of 128 (BW = 9.6 kHz), using a 2 kHz sinusoidal input sampled at 2.45 MHz.

Chapter 10 shows the design and implementation of a wideband SI A/D converter core suitable for undersampling and parallel time-interleaved operation. Spectral analysis on circuit-level simulation data and measurement results are included. A 20 MHz input bandwidth was achieved at a 3 MHz sampling rate. This is currently the *highest measured input bandwidth* for any SI ADC. The peak ENOB is 7.43-b, which is the second highest for any CMOS SI ADC with $f_{in} \geq 1$ MHz. A thorough comparison with previously reported SI ADC implementations is done using several different figures-of-merit.

Chapter 11 evaluates the potential for SI A/D converters in telecommunication applications. The ADC core in chapter 10 was used as a building block in a parallel time-interleaved ADC. By using eight parallel core ADCs, a sampling rate of 32 MS/s was achieved. This is currently the *highest sampling rate* for which measured performance has been reported for any SI ADC. The effects of channel mismatch are discussed, as they are seen to limit the performance, in particular the bandwidth.

Chapter 12 summarizes the conclusions drawn from the work presented in this book. The author also gives his view on the past and future evolution of the SI technique from a more industrial perspective.

Appendix

The **Appendix** has a collection of noise integrals that were used for the noise analysis in chapter 2.2.

REFERENCES

[1] C. Toumazou, F. J. Lidgey, D. G. Haigh, (Eds.) *Analogue IC design: the current-mode approach*, IEE Circuits and Systems series 2, 1990.

[2] C. Toumazou, J. B. Hughes, N. C. Battersby, (Eds.) *SWITCHED-CURRENTS an analogue technique for digital technology*, IEE Circuits and Systems series 5, 1993.

[3] S. J. Daubert, D. Vallancourt, and Y. P. Tsividis, "Current Copier Cells", *Electron. Lett.*, Vol. 24, No. 25, pp. 1560-1562, Dec. 1988.

[4] B. Jonsson, and S. Eriksson, "New Clock-Feedthrough Compensation Scheme for Switched-Current Circuits", *Electron. Lett.*, Vol. 29, No. 16, pp. 1446-1447, Aug. 1993.

PART I

AN INTRODUCTION TO THE SI TECHNIQUE

Chapter 2

Switched-Current Circuits

Basic functional primitives, such as summation, delay, sign inversion and scaling, are of fundamental importance in analog sampled-data signal processing systems. Data-conversion systems also include comparison and D/A conversion functions. This chapter describes the building blocks most commonly used to realize *switched-current* (*SI*) filters and data-conversion circuits. First their ideal behavior is explained in section 2.1. Then a number of non-ideal effects are discussed in section 2.2. The most significant problem in SI circuits is *charge injection* from MOS switches, also known as *clock-feedthrough* (*CFT*). A number of CFT compensation techniques are illustrated in section 2.3. Most of the performance of an SI circuit is set by the limitations of the *current sample-and-hold* (*CSH*). Section 2.4 closes the chapter with an overview of the evolution of the CSH or *SI memory cell* (*SI MC*).

2.1   Current-mode building blocks

2.1.1   The current mirror

In spite of its time-continuous nature, the current mirror well deserves to be mentioned as a building block for SI circuits. It is often used to implement signal scaling and filter coefficients in all kinds of SI filters and systems. Furthermore, its operation and circuit topology are similar to that of the SI memory cell. Thus, they have many common properties and problems. The *basic current mirror* is shown in Fig. 2.1 a. Its operation can be analyzed as follows: Assume that M0 and M1 are both operated in the saturation region. The gain, α, is realized by choosing the transistor width and length ratios according to $W_1/L_1 = \alpha(W_0/L_0)$. The drain currents for M0 and M1 can then be written as:

$$i_{D0} = I + i_{in} = \frac{\beta_0}{2}(v_{GS0} - V_T)^2(1 + \lambda v_{DS0}) \tag{2.1}$$

$$i_{D1} = \alpha I + i_{out} = \alpha \frac{\beta_0}{2}(v_{GS1} - V_T)^2(1 + \lambda v_{DS1}) \tag{2.2}$$

Assume that $v_{DS1} \approx v_{DS0}$ and/or that the channel length modulation λ is negligible. Obviously $v_{GS1} = v_{GS0}$, and thus, $i_{out} = \alpha \cdot i_{in}$. Note that the direction of i_{out} is reversed. Thus, the current mirror *always* performs a *sign inversion*. The case when the channel length modulation is *not* negligible is treated later in this chapter. It will of course introduce errors in i_{out}. Detailed descriptions of various current mirrors can be found in the literature, e.g., in [1].

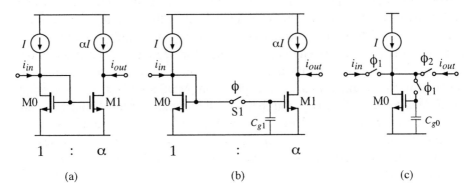

Fig. 2.1 Switched-current primitives. (a) Current mirror with current gain α. (b) First generation SI memory cell. (c) Second generation SI memory cell. MOS switches not explicitly drawn.

2.1.2 First and second generation SI memory cells

The first generation SI memory cell is shown in Fig. 2.1 b. It was presented in [2]. The only difference from the current mirror is the switch S1. The operation is controlled by a clock signal ϕ, and is described as follows. If the clock signal ϕ is high, then the switch transistor is on, $v_{GS1} = v_{GS0}$, and the circuit acts as a current mirror, i.e., the output *tracks* the input. Assume that ϕ goes low and the switch transistor turns off at the time $t = t_0$. The gate capacitance C_{g1} is sufficient to *hold* the gate-source voltage $v_{GS1}(t_0)$, and thus i_{out} is held at $i_{out}(t_0)$. The track-and-hold function is illustrated in Fig. 2.2 a.

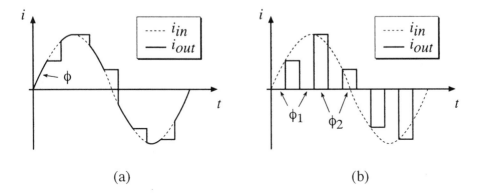

Fig. 2.2 Basic function of SI memories. (a) Track-and-hold cell
(b) Current copier cell

The second generation SI memory cell shown in Fig. 2.2 c was also presented in [2] as a basic building block for SI signal processing, and originally proposed in [3]. Three switches, controlled by a non-overlapping two-phase clock, are used to reconfigure the circuit to operate as both input and output circuit alternately. During phase 1, M0 is diode-connected just as M0 in Fig. 2.1 b, the output is disconnected and the circuit behaves like the input stage of the first generation SI memory. In phase 2, the diode-connection is broken and the input is disconnected. The gate-source voltage v_{GS0} is stored on the gate-capacitance C_{g0}, and thus the drain current i_{D1} is held constant and equal to $i_{D1}(t_0)$ if the switches are shifted at $t = t_0$. The output current will alternate between a valid sample and zero value, as shown in Fig. 2.2 b. It is obvious that this current memory cell can only have unity gain, since the same transistor M0 is used both as input and memory transistor. This leads to perfect matching between input and output circuits, since they are the same. Therefore, ideally, this circuit would realize a perfect unity gain memory cell. There are, however, other non-ideal effects that will introduce errors. They will be described later in this chapter. Much of the original work presented in this book require SI memory cells with multiple outputs, sometimes with different gain for each output. This originally lead the author to use first generation circuit building blocks. For that reason, the problem formulation and solutions in this book are biased toward first generation SI circuits, although many results are still applicable to second generation circuits. Throughout the rest of this book, the term "*SI memory cell*" or "*current sample-and-hold*" will denote a first generation SI memory, unless otherwise indicated.

2.1.3 OTA based current S/H circuits

The current mirror operation primarily relies on the transconductance of the MOS transistor to perform a I-to-V conversion directly followed by a V-to-I conversion. By replacing the mirror transistors with *Operational Transconductance Amplifiers* (OTAs), it is claimed that a more linear SI memory is realized [4]. A 20 MS/s A/D-converter with 7.7 effective number-of-bits resolution was implemented in BiCMOS using OTA-based SI circuits [5]. The feasibility of the technique was thereby shown, although part of the competitive performance may be due to the speed gained by using a BiCMOS process. An OTA-based SI memory cell realization is shown in Fig. 2.3. Its functionality is equivalent to the first-generation SI memory cell described previously. More detail can be found in [6].

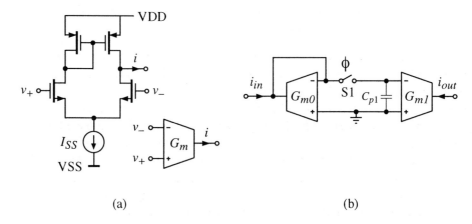

(a) (b)

Fig. 2.3 OTA-based SI memory cell. (a) OTA realization and symbol. (b) "First-generation" SI memory realization.

2.1.4 Comparators

Comparators are essential building blocks for A/D converters. In many ADC architectures, their performance imposes a direct limitation to the achievable resolution or speed. The most basic current comparator is shown in Fig. 2.4 a. It was used in the first current-mode ADC reported [7] and is still an alternative when low power dissipation and low offset is desired. Transistors M1 & M2 form a simple CMOS inverter. The input current is integrated by the gate parasitic capacitances C_{g1} and C_{g2}, and therefore there is no DC offset. Power dissipation is low since one MOST is usually off, except when close to the trip-point. The drawback is that the input voltage can have almost rail-to-rail swing and therefore the time it takes for a weak i_{in} to trip the comparator limits the speed. The current comparator proposed by Träff [8] is drawn in Fig. 2.4 b. The transistors M3 and M4 are used in a feedback configuration that limits the voltage swing both at the input and output nodes. Since the comparator is always close to its trip point its speed is increased. Another benefit is that the input impedance becomes very low. It can be shown that

$$R_{in} \approx \frac{1}{(g_{m3} + g_{m4})} \cdot \frac{(g_{ds1} + g_{ds2})}{(g_{m1} + g_{m2})} \qquad (2.3)$$

Power dissipation is increased since all of M1-4 are conducting DC current. Another drawback is that there is a small input offset current. Still, the Träff comparator represents a fast, yet simple, current comparator. A latch and two inverters are added in Fig. 2 b. The inverters restore logic levels for the digital output d, and the latch ensures a stable output for at least one half of a clock period. Several current comparators have been proposed in the literature, e.g., [9-13]. A more elaborate treatment of current comparators is found in [14].

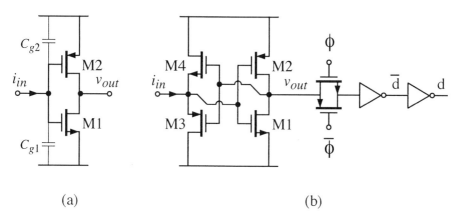

Fig. 2.4 Current comparators. (a) CMOS inverter. (b) Träff comparator.

2.1.5 D/A-converters

Current-mode D/A converters (DACs) can be used as stand-alone components or building blocks in A/D converters or programmable analog devices. A simple realization of a current-mode DAC is shown in Fig. 2.5. A number of binary weighted current sources are switched in and out as controlled by the digital word $d_0 \ldots d_{N-1}$. The resulting output current is described by Eq. 2.4.

$$i_{out} = I \sum_{j=0}^{N-1} d_j 2^{-j} \qquad (2.4)$$

A special case of the N-bit DAC is the 1-b DAC. In its basic form it is nothing more than a single current source controlled by a current-steering switch. It is frequently used in 1-bit Δ-Σ modulators and 1-b/stage ADCs.

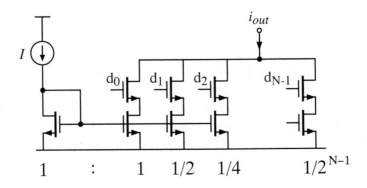

Fig. 2.5 Basic N-bit current D/A converter using binary weighted current sources.

2.2 Non-ideal effects

The main advantages pointed out for the SI technique has been its high-speed, low-voltage potential and its compatibility with digital CMOS technology. There is however a set of technical limitations associated with current-mode circuits. As the technique matures more work has to deal with analyzing and compensating for drawbacks rather than dwelling on the, sometimes unreachable, potentials. In a practical realization of SI circuits, non-ideal effects will always occur. Depending on the target system, these errors will have a different impact on the overall system performance. In some systems, e.g., in linear filters, a constant offset term can be considered a minor problem, while linear and non-linear errors may be disastrous. If a circuit is to perform logic comparison of a signal, a constant offset can be a severe problem, while gain errors and harmonic distortion may be of secondary importance. From a system designer's perspective, it is important to discern which errors are critical and should be eliminated. Non-ideal

Switched-current circuits 15

effects in SI circuits may show up as a constant offset, a linear gain error or a non-linear error, or a combination thereof.

2.2.1 MOS transistor mismatch

Chip fabrication inevitably introduces systematic and random errors to the circuit. Some are due to the variation of process parameters such as doping concentration. Other errors depend on the finite precision of device width and length. Although IC technology is continuously refined, fabrication-related errors will always occur, and the problems must be met by proper design.

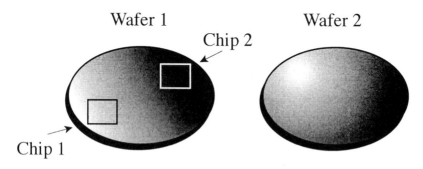

Fig. 2.6 Process parameter variation.

2.2.1.1 Process parameter variation

Process characterization data, such as parameter distributions, is normally provided from the chip foundry. Figure 2.6 shows how parameters like *doping concentration* and *oxide thickness* may vary over a single wafer. The ultimate effect is that V_T, K', etc., will have a *global* variation (different values on Chip 1 and Chip 2) and a *local* variation (on-chip gradient). Wafers 1 and 2 illustrate the difference that may occur between wafers, especially when originating from different runs. ***Global*** variation causes a level shift in parameter values and is primarily accounted for in the *circuit* design phase. Circuit performance is analyzed using *Monte Carlo* and *Worst Case Corner* analysis, allowing the device parameters to vary according to the random distributions supplied by the vendor. ***Local*** variation causes mismatch between devices that are close to each other and should be addressed in the layout design phase. *Common centroid* or *interdigitized* layout geometry can be used to improve matching between MOS devices [1].

2.2.1.2 Geometric layout errors

Due to mask errors, photolithography effects and variations in etching and lateral diffusion, the effective device dimensions, W_{eff} and L_{eff} are not equal to the drawn W and L, respectively. This effect is demonstrated in Fig. 2.7. W_{eff} and L_{eff} can be written as:

$$W_{eff} = W - 2WD + w$$
$$L_{eff} = L - 2LD + l \qquad (2.5)$$

where w and l are local random errors caused by the *finite edge resolution* on each device. LD is the *lateral diffusion* parameter causing a *systematic offset* in L. Similarly, WD will alter the device width [1].

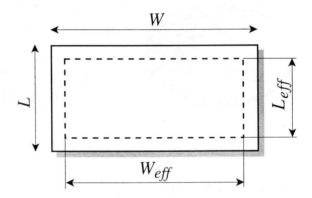

Fig. 2.7 Difference between *drawn* W/L and *effective* W/L for a MOS device.

2.2.1.3 Mismatch analysis

The following analysis is similar to that found in [15]. Consider the track-and-hold circuit in Fig. 2.1 b. The ideal relationship $i_{out} = \alpha \cdot i_{in}$ cannot be realized exactly, because of component mismatch. The mismatch is due to processing techniques and involves *physical parameters* such as V_T, as well as the *geometry of devices*, i.e., W and L. The three types of mismatches analyzed below are:
- β (transconductance) mismatch
- V_T (threshold voltage) mismatch
- λ (channel length modulation) mismatch

The transconductance parameter, $\beta = \mu C_{ox} W/L$, includes variations of μ and C_{ox}, as well as the aspect ratio. The channel length modulation and threshold voltage are physical parameters that change slowly over the chip. Let $\alpha = 1$ and assume that $v_{DS1} \approx v_{DSn}$ and held approximately constant. This is true for most circuits practically used. Due to mismatch, $\beta_1 = \beta_0 + \Delta\beta$, $V_{T1} = V_{T0} + \Delta V_T$, and $\lambda_1 = \lambda_0 + \Delta\lambda$. Then, Eqs. 2.1 and 2.2 can be rewritten as:

$$i_{D0} = I + i_{in} = \frac{\beta_0}{2}(v_{GS} - V_{T0})^2(1 + \lambda_0 v_{DS0}) \tag{2.6}$$

$$i_{D1} = I + i_{out} = \frac{\beta_0 + \Delta\beta}{2}\left(v_{GS} - (V_{T0} + \Delta V_T)\right)^2\left(1 + (\lambda_0 + \Delta\lambda)v_{DS0}\right) \tag{2.7}$$

Then we get:

$$(1 + \varepsilon) = \frac{i_{D1}}{i_{D0}} =$$

$$= \left(1 + \frac{\Delta\beta}{\beta_0}\right)\left(1 + \left[\frac{\Delta V_T}{v_{GS} - V_{T0}}\right]^2 - 2\frac{\Delta V_T}{v_{GS} - V_{T0}}\right)\left(1 + \frac{\Delta\lambda v_{DS0}}{1 + \lambda v_{DS0}}\right) \tag{2.8}$$

where ε is the mismatch error. Since $i_{out} = i_{D1} - I$, we also get

$$i_{out} = (1 + \varepsilon)i_{in} + \varepsilon I \tag{2.9}$$

Thus, the mismatch will cause gain *and* offset errors. It is seen from Eq. 2.8 that the β and λ mismatch terms are signal independent and cause only *linear gain* and *constant offset* errors. In order to study the V_T mismatch more closely, let $\Delta\beta = 0$ and $\Delta\lambda = 0$. From Eqs. 2.6 and 2.7 we get that

$$(v_{GS} - V_{T0})^2 \approx \frac{2}{\beta_0}i_{D0} \tag{2.10}$$

$$i_{D1} \approx i_{D0} + \frac{\beta_0}{2}\Delta V_T^2 - \sqrt{2\beta_0 i_{D0}}\,\Delta V_T \tag{2.11}$$

We can see that V_T mismatch will cause *constant offset* and *non-linear gain* errors. Any mismatch error, e, in the bias current source, such that it becomes $I\cdot(1 + e)$, will only cause a constant offset $-e\cdot I$ in i_{out}.

2.2.2 Conductance ratio errors

Ideally, a current mirror should have infinite output impedance and zero input impedance. In reality, the output impedance of the simplified current-mode circuits shown in Fig. 2.8 is $1/g_{ds1}$ and the input impedance is $1/g_{m2}$. The resulting current transfer is

$$\frac{i_{d2}}{i_{d1}} = \frac{g_{m2}}{g_{m2} + g_{ds1}} = \frac{1}{1 + \frac{g_{ds1}}{g_{m2}}} \approx 1 - \frac{g_{ds1}}{g_{m2}} = 1 + \varepsilon \qquad (2.12)$$

where

$$\varepsilon = -\frac{g_{ds1}}{g_{m2}} = -\frac{\lambda i_{D1}}{\sqrt{2K' i_{D2} W/L}} = -\frac{\lambda}{\sqrt{2K' W/L}} \frac{I+i}{\sqrt{I-i}} \qquad (2.13)$$

It is seen from Eq. 2.13 that the error is strongly dependent on the signal i, and will therefore produce *non-linear distortion*. The transfer from the output transistor of one stage to the input transistor of another is *always* less than unity, as seen from Eq. 2.12. This indicates that a fraction of the signal will always be lost. In SI filters this will cause a pole/zero displacement, and in A/D converters there will be gain errors. Therefore SI circuits are usually implemented with some compensation technique that improves the conductance ratios. *Cascode* techniques [2, 16-20] can be used to increase output impedance while *active mirrors* [21-22] use active feedback to further decrease the input impedance. Implementations presented in this book use simple low-voltage cascodes [2, 23].

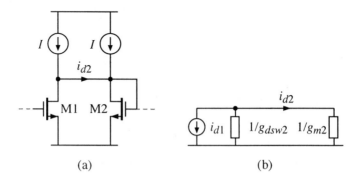

Fig. 2.8 Conductance ratio error.

2.2.3 Clock-feedthrough

Charge-injection or *clock-feedthrough* (CFT) is inherent in any sampled-data system using MOS switches. In SI circuits, CFT becomes one of the main performance limitations. The gate capacitance at the storage node is usually small and therefore even a small amount of charge will alter the stored current significantly. It is also a drawback that the relation between injected charge and output current error is non-linear. Although the mechanisms behind charge injection may be well understood, they are difficult to describe in closed form expressions. Still there is a need for models that capture the main sources of CFT. Most publications dealing with the CFT problem in SI circuits, e.g., [24-28], seem to favor model simplicity, rather than in-detail accuracy. In [29], however, the analysis is more in line with general results published on charge injection in MOS switches, e.g., [30-31]. A detailed résumé of models used to described channel charge redistribution and clock signal coupling through overlap capacitances was given by the author in his Licentiate thesis *"Applications of the Switched-Current Technique"* [32]. In this section a more summarized treatment of CFT modeling is given.

2.2.3.1 Clock-feedthrough parasitics

The source of CFT is the coupling of the clock signal onto the memory transistor gate through parasitic capacitances. These parasitics can be referred to the gate-source overlap formed by the *lateral diffusion* of the drain and source areas, or to the gate-channel capacitance. Fringing field parasitics will also contribute to the effect, but are not included in the analysis. Fig. 2.9 a shows an MOS transistor cross section with the gate overlap indicated. The area of the overlap capacitance is $W_{eff}LD$, and thus:

$$C_{ovl} = C_{ox} W_{eff} LD \qquad (2.14)$$

where LD is the lateral diffusion length, $W_{eff} = (W - 2WD)$ is the effective channel length, and C_{ox} is the thin-oxide capacitance per unit area. WD is the field-oxide encroachment parameter which reduces W as LD reduces L.

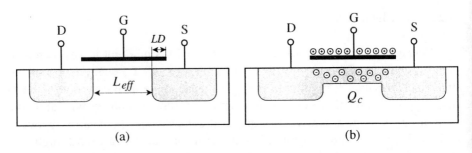

Fig. 2.9 MOS transistor cross section. (a) The channel in the cut-off region, and the gate overlap capacitance caused by lateral diffusion. (b) The channel charge, and the channel in the linear region.

In Fig. 2.9 b, the gate-channel capacitance (C_{gc}) is visible. The transistor channel forms the bottom plate of C_{gc}. Apparently there is no distinct boundary between C_{gc} and C_{ovl} since they are adjacent, but C_{gc} is considered to cover the effective gate area, $W_{eff}L_{eff}$. The channel charge, Q_c, is associated with that area. We know that the channel starts to build up from charges induced by V_G when $V_{GS} > V_T$. Consequently, the channel charge is

$$Q_c = (V_{GS} - V_T) \cdot C_{gc} \quad \text{where} \tag{2.15}$$

$$C_{gc} = C_{ox} W_{eff} L_{eff} = C_{ox} W_{eff} (L - 2LD) \tag{2.16}$$

for $V_{GS} > V_T$. As seen from the cross section, C_{gc} is electrically connected with both drain and source in a *distributed* fashion through the channel. When the gate-source voltage drops below V_T, the conducting channel disappears and Q_c is forced to redistribute onto the drain and source terminals. The amount of charge injected into the drain and source can only be calculated using numerical methods [30-31, 33]. It depends strongly upon the relative speed of signal transients, meaning that, for an MOS switch, *the slope* of the falling clock edge is an important design parameter. Closed form expressions that can be derived for special cases are included in the summary below.

Switched-current circuits

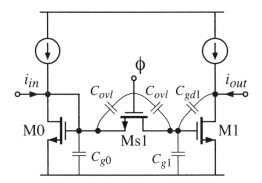

Fig. 2.10 Switched-current memory cell with parasitic capacitances.

An SI memory circuit with parasitic capacitances included is shown in Fig. 2.10. When ϕ is greater than $V_T + v_{G0}$, the switch is conducting with a conductance $g(\phi)$, and when ϕ is less than $V_T + v_{G0}$, the switch has zero conductance. This leads to the two small signal equivalent circuits shown in Fig. 2.11 a and b. The distributed nature of the switch is indicated by the segmentation of C_{gc} in Fig. 2.11 a. If M0 and M1 are designed to operate in the saturated region, we find from [1] that:

$$C_{gi} = \frac{2}{3}C_{gci} + C_{ovli} = C_{ox}W_{eff,i}\left(\frac{2}{3}L_{eff,i} + LD\right) \quad i = 0,1 \tag{2.17}$$

$$C_{gd,i} = C_{ovl,i} = C_{ox}W_{eff,i}LD \quad i = 1 \tag{2.18}$$

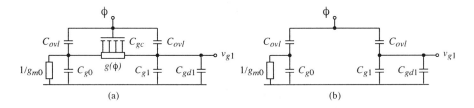

Fig. 2.11 SI memory cell small signal equivalent circuit for (a) $\phi - v_{G0} > V_T$. Capacitances involved in clock-feedthrough are shown, and the distributed nature of the switch channel is indicated. (b) $\phi - v_{G0} < V_T$. The switch is *off* and there is no interaction between drain and source.

2.2.3.2 Influence of clock slope

The correct modeling of the distributed channel capacitance depends on how steep the clock edges are. A detailed analysis of charge injection in MOS switches is found in [30-31]. The distributed and dynamic nature of the switch transistor leads to a set of differential equations that requires numerical solutions. A switching parameter, B, determining the *relative switching speed* was defined in [30] as

$$B = (v_{GS,s1} - V_{Ts1})\sqrt{\frac{\beta_{s1}}{a\,C_{g1}}} \approx (v_{GS,s1} - V_{Ts1})\sqrt{\frac{3\mu_0\,W_{s1}/L_{s1}}{2a\,W_1 L_1}} \qquad (2.19)$$

where

$$a = \frac{\Delta\phi}{t_f} = \frac{\phi_{high} - \phi_{low}}{t_f} \qquad (2.20)$$

is the slope of the clock edge, and t_f is the clock signal fall time. It is used to characterize *the first part of the clock transition*, when the switch is conducting, in the following way.

- If $B \gg 1$, the clock transition is said to be **slow**. Drain and source are connected through $g(\phi)$ and there is enough time for the terminals to communicate. This tends to keep the final drain and source voltages equal, and thus the charge partition is *proportional to the node capacitance* at each terminal. The small signal equivalent is shown in Fig. 2.12 a.
- If $B \ll 1$ it is considered **fast**. The conducting channel disappears almost instantly. There is not enough time for the charge at the drain side to communicate with the charge at the source side, and Q_c will split in *two equal parts*, no matter what the terminal load capacitances are. This corresponds to the circuit shown in Fig. 2.12 b.
- If $B \approx 1$, an **intermediate** case occurs which requires a numerical solution to the channel charge redistribution. The channel charge partition is then strongly depending on B. Unfortunately, $B \approx 1$ corresponds to many realistic situations. Note that B is *a function of several different kinds of design parameters*. Therefore, even the choice of a proper CFT model in this case depends on process parameters and device geometry, as well as on node voltages and signal rise times found in the circuit.
- If $C_{g1} = C_{g0}$, or rather $C_{g1} + C_{gd1} = C_{g0}$, the channel charge *always splits equally*, independent of B.

The last part of the transition occurs when $v_{GS,s1} < V_{Ts1}$. Then there is no conducting channel and therefore no interaction between source and drain, as shown in Fig. 2.11 b. Only the switch overlap capacitances inject charge onto the memory transistor gate.

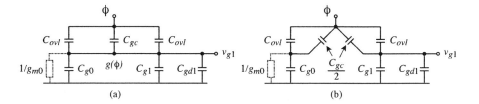

Fig. 2.12 SI memory cell small signal equivalent circuit for (a) *slow* clock transition. (b) *fast* clock transition.

case	CFT model
slow	$V_{CFT} \approx \Delta\phi \dfrac{3}{2}\left[\eta\dfrac{W_{s1}L_{s1}}{W_0L_0 + W_1L_1} + (1-\eta)\dfrac{W_{s1}LD}{W_1L_1}\right]$
medium	Numerical solution is required
fast	$V_{CFT} \cong \dfrac{3}{2} \cdot \dfrac{\Delta\phi W_{s1}\left(\dfrac{\eta}{2}L_{s1} + LD\right)}{W_1L_1}$
equal D & S load	$V_{CFT} \cong \dfrac{3}{2} \cdot \dfrac{\Delta\phi W_{s1}\left(\dfrac{\eta}{2}L_{s1} + LD\right)}{W_1L_1}$

Table 2.1 Clock-feedthrough models for different switch transition speed

2.2.3.3 Summary of CFT models

Closed-form models can be derived from the small signal equivalent circuits in Fig. 2.12. This was done in [32]. A summary of such simplified CFT models is given in Table 2.1, where the parameter

$$\eta = \frac{\left(\phi_{high} - v_{G0} - V_{Ts1}\right)}{\left(\phi_{high} - \phi_{low}\right)} \qquad (2.21)$$

denotes the fraction of the voltage swing that the switch is on. It was also shown by the author in [32] that many realistic circuits have a "medium" switching speed, which requires numerical solution of the CFT equations. In order to make quick estimations or analytical derivations, it will however be necessary to assume one of the special cases given below. Many authors have assumed "fast" switching, since most circuits are better described by that model than by the model for "slow" switching [32].

2.2.4 Noise

Circuit noise is essentially an analog, time-continuous, phenomenon. However, in analog sampled-data circuits, the circuit noise is picked up in the sampling process and added to the discrete-time data sequence. In this section, the effective noise power in the sampled-data sequence, resulting from time-continuous circuit noise will be estimated. Shaping of the noise power spectral density (PSD) function $S(f)$ will not be elaborated upon. The noise PSD *is* shaped by the sampling process, since all high-frequency noise-components are folded into the Nyquist band. Still the total noise-power remains unchanged.

2.2.4.1 MOS transistor noise

Noise in MOS transistors has been treated and extensively described in the literature, e.g., [34-35]. Noise in SI and current-mode circuits has also been analyzed in several publications, e.g., [6, 15, 36-42]. The discussion below will include some of the most fundamental relations. The current noise *power spectral density* (PSD) of *thermal, flicker* and *shot* noise, is described by Eqs. 2.22 a-c respectively.

$$S_{th}(f) = \frac{8kTg_m}{3} \qquad (2.22\text{ a})$$

$$S_{fl}(f) = \frac{K_F I_{DQ}^{A_F}}{C_{ox} WL} \cdot \frac{1}{f} \qquad (2.22\text{ b})$$

$$S_{sh}(f) = 2qI_{dj} \qquad (2.22\text{ c})$$

K_F and A_F are the flicker noise constants, and I_{dj} is the junction current in the parasitic diodes associated with the MOS transistor. These diodes are normally reverse-biased, and therefore the shot noise component is small compared to thermal noise. If flicker noise is a problem, it can be eliminated by correlated double sampling, which is inherent in second-generation SI circuits [43]. For the wideband applications found in chapters 10 and 11 in this book thermal noise is the dominating problem and therefore the following noise calculations will focus on thermal noise. A MOS transistor operating in the linear region is modeled as a thermal noise source, identical to a passive resistor having $R = 1/g_{ds}$. Its noise current density becomes

$$S_{th}(f) = 4kTg_{ds} \tag{2.23}$$

Shot noise and thermal noise has a constant PSD for all frequencies. In practice, high-frequency noise components will be attenuated according to the transfer function $H(f)$ in the analog signal path from noise source to observation or sampling point. The resulting mean-square noise current can be written as

$$\overline{I_w^2} = S_w \int_0^\infty |H(f)|^2 df \tag{2.24}$$

where S_w is the (constant) white-noise spectral density. For a single-pole system with 3 dB cutoff frequency f_0, it can be shown that

$$\int_0^\infty |H(f)|^2 df = \int_0^\infty \left| \frac{1}{1 + j(f/f_0)} \right|^2 df = f_0 \frac{\pi}{2} = \frac{p_0}{4} \tag{2.25}$$

and therefore the total *root-mean-square* (rms) noise current becomes

$$\sqrt{\overline{I_w^2}} = \sqrt{S_w \frac{p_0}{4}} \tag{2.26}$$

where $p_0 = 2\pi f_0$. A few similar integrals for relevant transfer functions are compiled in "*Appendix – Noise Integrals*". The results are used in the following analysis.

2.2.4.2 Sampled noise

A simplified first-generation SI circuit is shown in Fig. 2.13. It consists of two sample-and-hold circuits forming a single clock cycle delay. The noise contributions from each of the transistors M0-M3, MP0-MP3 are shown as noise current sources. Ideally, the drain currents $i_{d1}(t)$ and $i_{d2}(t)$ in their hold phases represent the sampled-data sequence $i_{d1}(k)$ and $i_{d2}(k)$, where $i_{d1}(k) = i_{in}(k)$. The input signal $i_{in}(t)$ will however be contaminated by any circuit noise affecting node 'A', where sampling occurs. Noise sources i_{nP0}, i_{n0} and i_{ns1} cause a noise voltage on 'A' resulting in a noise current $i_{e1}(t)$ at the drain of M1. After sampling at 'A' this noise is represented in the sampled data sequence as $i_{e1}(k)$ so that

$$i_{d1}(k) = i_{in}(k) + i_{e1}(k) \tag{2.27}$$

where $i_{e1}(k)$ has the same rms-value as $i_{e1}(t)$. From node 'A' to node 'B', additional circuit noise will contaminate the analog signal and be picked up by the sampling operation at 'B'. The sampled-data sequence becomes

$$i_{d2}(k) = i_{d1}(k) + i_{e2}(k) = i_{in}(k) + i_{e1}(k) + i_{e2}(k) \tag{2.28}$$

where $i_{e2}(k)$ is the sampled noise contribution from M1-2, MP1-2 and Ms2. In general, after N stages, the sampled value is

$$i_{dN}(k) = i_{in}(k) + i_{e1}(k) + i_{e2}(k) + \ldots + i_{eN}(k) = i_{in}(k) + \sum_{j=1}^{N} i_{ej}(k) \tag{2.29}$$

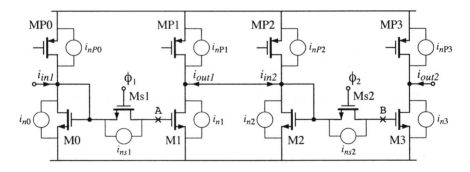

Fig. 2.13 Noise sources affecting the sampled-data sequence in a simplified first generation SI circuit.

Switched-current circuits often have unity, or close to unity gain. For N identical unity-gain stages, the rms noise current is approximately proportional to the square-root of N. In order to estimate the white-noise power added by a single stage, we consider the noise added from 'A' to 'B', or rather in the signal path from the sampled i_{d1} to i_{d3}. The small-signal equivalent circuit is shown in Fig. 2.14, where it has been assumed that $g_{ds} \ll g_m$ where relevant. Noise from transistors M1-M2, and MP1-MP2 is represented by a single noise source i_{nB}.

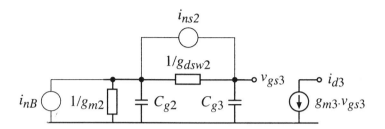

Fig. 2.14 Small-signal equivalent schematics including circuit noise sources.

The transfer functions from the two noise sources to i_{d3} are

$$H_{nB}(s) = \frac{g_{m3}}{g_{m2}} \cdot \frac{1}{\left(1 + s\left[\dfrac{C_{g2} + C_{g3}}{g_{m2}} + \dfrac{C_{g3}}{g_{dsw2}}\right] + s^2 \dfrac{C_{g2} \cdot C_{g3}}{g_{m2} \cdot g_{dsw2}}\right)} \quad (2.30)$$

$$H_{ns2}(s) = \frac{g_{m3}}{g_{dsw2}} \cdot \frac{\left(1 + s\dfrac{C_{g2}}{g_{m2}}\right)}{\left(1 + s\left[\dfrac{C_{g2} + C_{g3}}{g_{m2}} + \dfrac{C_{g3}}{g_{dsw2}}\right] + s^2 \dfrac{C_{g2} \cdot C_{g3}}{g_{m2} \cdot g_{dsw2}}\right)} \quad (2.31)$$

Thus, the resulting mean-square noise current is

$$\overline{i_n^2} = S_{nB} \cdot \int_0^\infty |H_{nB}(f)|^2 df + S_{ns2} \cdot \int_0^\infty |H_{ns2}(f)|^2 df \quad (2.32)$$

where

$$S_{nB} = \frac{8}{3} kT \left(g_{m2} + g_{m3} + g_{mP2} + g_{mP3}\right) \quad (2.33)$$

$$S_{ns2} = 4kT g_{ds2} \quad (2.34)$$

$$\int_0^\infty |H_{nB}(f)|^2 df = \left(\frac{g_{m3}}{g_{m2}}\right)^2 \frac{p_0 p_1}{p_0 + p_1} \cdot \frac{1}{4} \tag{2.35}$$

$$\int_0^\infty |H_{ns2}(f)|^2 df = \left(\frac{g_{m3}}{g_{dsw2}}\right)^2 \left(1 + \frac{p_0 p_1}{z_0^2}\right) \frac{p_0 p_1}{p_0 + p_1} \cdot \frac{1}{4} \tag{2.36}$$

From Eq. 2.31 we know that

$$p_0 p_1 = \frac{g_{m2} \cdot g_{dsw2}}{C_{g2} \cdot C_{g3}} \tag{2.37}$$

$$p_0 + p_1 = \frac{g_{m2}}{C_{g2}} + g_{dsw2}\left(\frac{1}{C_{g2}} + \frac{1}{C_{g3}}\right) \tag{2.38}$$

$$z_0 = \frac{g_{m2}}{C_{g2}} \tag{2.39}$$

Under the assumption that M3 = M2, MP3 = MP2, the total mean-square current noise sampled into M3 is

$$\begin{aligned}\overline{i_n^2} &= kT\left[\frac{4}{3}(g_{m2} + g_{mP2}) + g_{dsw2}\left(\frac{g_{m3}}{g_{dsw2}}\right)^2\left(1 + \frac{p_0 p_1}{z_0^2}\right)\right]\frac{p_0 p_1}{p_0 + p_1} = \\ &= kT\left[\frac{4}{3}(g_{m2} + g_{mP2}) + g_{m2}\left(1 + \frac{g_{m2}}{g_{dsw2}}\right)\right]\left(\frac{g_{m2}}{C_{g2}} \cdot \frac{g_{dsw2}}{g_{m2} + 2g_{dsw2}}\right)\end{aligned} \tag{2.40}$$

and the signal-to-noise ratio (SNR) for a sinusoidal input current with magnitude A becomes

$$SNR = 10\log\left\{\frac{\overline{i_{in}^2}}{\overline{i_n^2}}\right\} =$$

$$= 10\log\left\{\frac{\frac{A^2}{2}}{kT\left[\frac{4}{3}(g_{m2}+g_{mP2})+g_{m2}\left(1+\frac{g_{m2}}{g_{dsw2}}\right)\right]\left(\frac{g_{m2}}{C_{g2}}\cdot\frac{g_{dsw2}}{g_{m2}+2g_{dsw2}}\right)}\right\} \quad (2.41)$$

Clearly, SNR is improved if g_{m2}, g_{mP2} and g_{dsw2} are chosen as small as possible. Unfortunately that will also reduce the bandwidth, unless C_{g2} & C_{g3} are reduced at the same time. This, on the other hand, would increase clock-feedthrough and matching errors. A low g_{m2} also increases the voltage swing at the input node. For low voltage systems, there is not much freedom in choosing these voltage levels. Careful design tradeoffs are required in high-speed, high-performance, low-voltage SI system design. Additional noise is sampled if a number of current mirrors precede the SI sampling. A pessimistic estimate of the noise contribution of such mirrors is achieved by repeating the above analysis with the switch shorted. Without the switch, the current mirror becomes a single-pole system, and the equivalent of Eq. 2.40 and 2.41 is

$$\overline{i_n^2} = kT\frac{4}{3}(g_{m2}+g_{mP2})p_0 = kT\frac{4}{3}(g_{m2}+g_{mP2})\frac{g_{m2}}{2C_{g2}} \quad (2.42)$$

$$SNR = 10\log\left\{\frac{\overline{i_{in}^2}}{\overline{i_n^2}}\right\} = 10\log\left\{\frac{\frac{A^2}{2}}{kT\frac{4}{3}(g_{m2}+g_{mP2})\frac{g_{m2}}{2C_{g2}}}\right\} \quad (2.43)$$

This SNR estimate is pessimistic because the noise power is calculated over an infinite bandwidth with a single-pole high frequency roll-off. In reality the high-frequency noise in any preceding time-continuous circuits is further attenuated by all subsequent circuits between it and the sampling gate. Still, Eq. 2.42 and 2.43 are useful for quick estimations.

2.2.5 Settling

Settling errors for second-generation SI memory cells have been treated in the literature, e.g., [15]. The analysis below focuses on the settling behavior of the first-generation SI memory. Consider the circuit in Fig. 2.13, and the small signal equivalent in Fig. 2.14. The current transfer function from i_{in2} to i_{out2} is given by Eq. 2.30. For a unity-gain memory cell we assume that

$$C_{g3} = C_{g2}$$
$$g_{dsw2} = x \cdot g_{m2}$$
(2.44)

Combining Eqs. 2.37, 2.38 and 2.44 gives the two real poles p_0 and p_1.

$$\{p_0, p_1\} = \frac{g_{m2}}{2C_{g2}}\left(1 + 2x \mp \sqrt{1 + 4x^2}\right)$$
(2.45)

The exact value of x depends on the operating point of the memory cell, and the geometry of the switch and memory transistors. With the transistor sizes and operating point used for the ADC design in chapters 10 and 11, x is approximately 0.2. Typical values would range from 0.1 to about 1. The values of poles p_0 and p_1 (in units of $g_{m2}/2C_{g2}$) is plotted in Fig. 2.15 for x ranging from 0.1 to 10. For high x values, p_0 approaches the single pole of a current-mirror with identical design. At low x values, the settling is excessively slowed down by the poor switch conductance. As a rough approximation of the settling error, we can assume that the memory cell has a single-pole behavior set by p_0. With that assumption, the settling error for a step Δi_{in} in the input current is described by Eq. 2.46.

$$\varepsilon(t) = e^{-p_0 t} \Delta i_{in}$$
(2.46)

Linear small signal models were used in the above analysis. In reality, g_{m2} is a non-linear function of i_{in}. It will change from about +22 to -29% of its nominal value if the input goes from +50 to -50% of the bias current. Settling errors in SI circuits should therefore generally be considered as non-linear errors, and circuit-level simulations are necessary for accurate design.

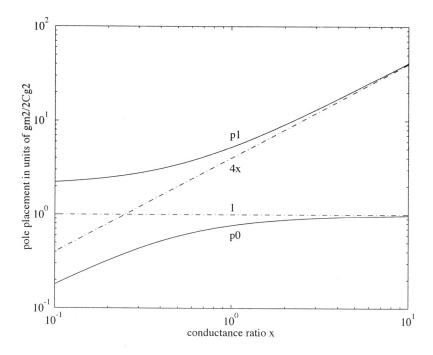

Fig. 2.15 Pole placement vs. relative switch conductance.

2.2.6 Voltage drop

In practical implementations of SI circuits there will always be a certain amount of resistive voltage drop on the on-chip power supply wires. These voltage drops will impair drain current matching, and it can be shown that a careful design of wire resistance may be necessary to be able to achieve good current matching. This was treated in detail by the author in the paper *"Design of power supply lines in high-performance SI and current-mode circuits"* [44] which is included in this book. (See chapter 6)

2.2.7 Jitter

An accurately defined sampling time interval is essential in any sampled-data system. Sampling time variations, or *jitter*, at the input sample-and-hold inevitably produces non-recoverable errors in the signal. It can be shown that the higher the input frequency, the more severe the jitter effects [45]. Sampling jitter is a major performance limitation in high-speed, high-accuracy, sampled-data systems. Sampling time uncertainty can be random or signal dependent. Random jitter is the result of noise on the clock signal, and will appear as noise in the output signal. Signal dependent jitter is due to signal dependent switch-off time of the sampling switch. It will add harmonic distortion to the signal. Sampling jitter is treated by the author in the papers *"Distortion in Sampling"* [46], and *"Sampling Jitter in High-Speed SI Circuits"* [47] of which the latter is included in this book. (See chapter 5)

2.2.8 Summary of non-ideal effects

The different types of errors produced by different non-ideal effects are summarized in Table 2.2. Note that many non-linear errors can be separated into a linear and a non-linear component, and sometimes also a constant offset term.

Source of error	Offset	Linear	Non-linear	Noise
Device geometry	•	•		
C_{ox}, μ, λ	•	•		
V_T	•	•	•	
Conductance ratio		•	•	
Clock feedthrough	•	•	•	
I_{bias} mismatch	•			
voltage drop	•	•	•	
sampling jitter			•	•
finite settling		•	•	
noise				•

Table 2.2 Summary of non-ideal effects and the resulting errors.

2.3 Clock-feedthrough compensation

Since CFT is the most severe limitation on SI circuit performance, a large number of compensation schemes have been proposed in the literature. Most of the suggested approaches can be subdivided into six main groups: *attenuation, cancellation, algorithmic, feedforward/feedback, adaptive* and *zero-voltage switching* techniques. The author's contribution in this field is a low-CFT SI circuit of cancellation type [48-50]. It is described in detail in chapter 4, *"Clock-Feedthrough Compensated First-Generation Circuit Design"*. A short résumé of the six techniques is given below.

2.3.1 Attenuation techniques

Attenuation techniques aim to directly reduce the error voltage or the amount of charge that is actually injected onto the gate of the memory transistor. The most obvious way to achieve this is to increase the capacitance at the storage node. Unfortunately, the amount of capacitance needed for high attenuation is large, so this technique is only useful for low-speed applications. An improvement, shown in Fig 2.16 a, was included in [3]. The idea is to load the sampling switch with a large capacitance only at the charge-injection moment. As S1a turns off, S1b is still on, creating a negative feedback loop. The effective capacitance seen by S1a at switch-off is $C_3 + (1+T)C_M$, where T is the loop gain, and the charge injection from S1a becomes negligible. When S1b turns off, the feedback loop is broken, and the charge injected is split between C_1, C_M and C_3. If C_M is small compared to C_1, a noticeable CFT attenuation is achieved. In the sampling (tracking) phase, C_M is shorted by S1a and S1b, so the settling time is determined by $C_1 + C_3$. A similar circuit, with the voltage-follower excluded, was proposed in [51].

A slightly different approach was used in [52-53] where a separate Miller-capacitance is connected as a load to the hold node. Its feedback loop does not include the actual memory cell, thus enabling independent design. The loop gain around the Miller-capacitance is low during sampling (tracking) but switched to "high" just before the track-to-hold transition so that the CFT voltage is attenuated. A differential approach was used in [53] for both single-ended and fully-differential memory cells.

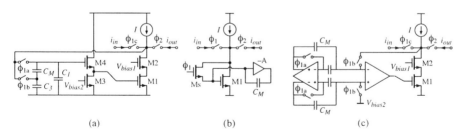

Fig. 2.16 CFT attenuation techniques. (a) Negative feedback (Miller) technique proposed in [3]. (b) Miller-enhancement technique in [52]. (c) Differential Miller technique in [53].

2.3.2   Cancellation techniques

Cancellation techniques combine two or more CFT error components in a controlled way to achieve total or partial cancellation. The success of cancellation techniques is strongly dependent on how accurately one can predict the included CFT components. A number of cancellation techniques have been proposed for SI circuits. One example is the *dummy switch* [25] which is already well known from switched capacitor and other voltage-mode circuits. Related to the dummy switch is the CMOS switch, where the designer can match the channel charges from the P- and NMOS transistors to cancel each other. Both CMOS and dummy switch techniques are sensitive to clock edge slope and timing variations, and in the case of a CMOS switch there is also the difficulty of matching process parameters as well as geometry. Still, the dummy switch has been used extensively in SI circuits found in the literature, sometimes in combination with other techniques. In [54] the dummy switch technique was combined with a common-gate amplifier that was used to keep the input voltage nearly constant. A kind of dummy-switch variation was proposed in [55], where a second switch in combination with a capacitor is used to balance the charge injected from the actual sampling switch.

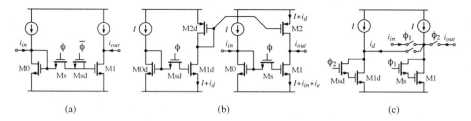

Fig. 2.17 CFT cancellation techniques. (a) Dummy switch. (b) Constant CFT cancellation in 1st generation MC. (c) Constant CFT cancellation in 2nd generation MC.

While the dummy switch aims to cancel the charge injected onto the memory transistor gate, other techniques allow the error to propagate into the current domain, where two or more error currents are combined in order to cancel each other. It was shown previously that a large part of the CFT is constant, although in many applications it is signal dependent errors, particularly harmonic distortion, that are the most undesired. *Constant* CFT errors can be removed with the help of a zero-input *dummy circuit* [25], as shown in Fig. 2.17 b. The dummy circuit produces a replica of the constant CFT error, i_d, which is inverted and copied to the output of the memory cell through Md2 – M2. Complete CFT cancellation is only achieved for $i_{in} = 0$, when $i_d = i_e$. For non-zero input signals the signal dependent part of i_e still remains, yet a considerable reduction in CFT error *magnitude* is often achieved. Later, the technique has been used to remove constant CFT in 2nd generation SI circuits [56], as well as S^{2}I [57] and S^{n}I circuits [58-60]. Figure 2.17 c shows a 2nd generation current memory with dummy circuit cancellation. The dummy error i_d is subtracted from i_{in} during the *input* phase, but the effect is essentially the same as in [25].

Switched-current circuits

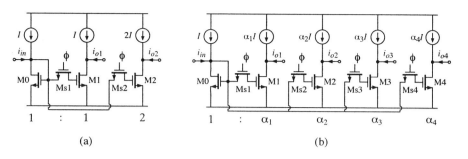

Fig. 2.18 CFT cancellation. (a) Signal-dependent CFT cancellation [26, 61]. (b) *Author's contribution*: 1st generation MC with complete CFT cancellation presented in [48, 50].

Signal dependent CFT errors are canceled with the circuit shown in Fig. 2.18 a. It was proposed in both [26] and [61] independently. The output current should be taken as $i_{out} = i_{o2} - i_{o1}$. If Ms1 is identical to Ms2 and the current gain of two is realized by choosing $W_2 = 2W_1$, then the gate capacitance of M2 is twice that of M1. Thus, the CFT error voltage on M2 becomes half as much as on M1. A small-signal analysis indicates that, since $g_{m2} = 2g_{m1}$, the CFT **current** errors are identical, and therefore cancel as the $i_{out} = i_{o2} - i_{o1}$ is established. Unfortunately, a large-signal analysis [26, 50] reveals that a small constant CFT residue still remains. A variation on the approach in [26] and [50] was proposed in [62], where also the PMOS current sources are sampled. Another variation is found in [63], where a diode-connected NMOS transistor replaces the PMOS current source in order to achieve a linear I/V-V/I conversion.

The earliest reported first generation SI memory cell with simultaneous cancellation of both constant *and* signal-dependent CFT errors was proposed by the author in [48]. The circuit realization is shown in Fig. 2.18 b. It constitutes an extension and generalization of the circuit proposed by Fiez/Song et. al. A *double* difference $i_{out} = (i_{o4} - i_{o3}) + (i_{o2} - i_{o1})$ is forming the output current and, by proper selection of $\{\alpha_1 \ldots \alpha_4\}$, complete CFT cancellation is achieved. In [50] it was shown how the coefficients can be optimized for speed, area, power or any other goal function. The only condition that has to be met is

$$\left(\frac{1}{\alpha_4} - \frac{1}{\alpha_3}\right) + \left(\frac{1}{\alpha_2} - \frac{1}{\alpha_1}\right) = 0 \tag{2.47}$$

A partially similar approach was used in [64] for second generation circuits, where instead the *switch size* is varied in three signal paths that are scaled and combined to form the final output. As in [50], it is shown that there is an infinite number of solutions that gives complete CFT cancellation. In [65] it was shown that complete CFT cancellation can also be achieved by a combination of the "dummy-circuit" technique and a circuit similar to that of Fiez/Song et. al. Lately, an error cancellation approach that is probably best characterized as a cancellation technique was proposed by Hughes et. al. in [66]. The *Error Neutralization* technique aims to reduce the effect of gate-drain parasitic capacitance in differential SI memory cells by the use of cross-coupled neutralization capacitances. It was also shown in [66] that its error reduction can be extended to include output conductance and CFT errors as well.

2.3.3 Algorithmic techniques

Algorithmic techniques all use some form of *iterative* sampling method where CFT errors are gradually canceled. The first memory cell of this type is the algorithmic current memory [18], later followed by the S^2I [67] and S^nI circuits [68] which are shown in Fig. 2.19. A two step algorithm is used in the S^2I memory cell. During phase $\emptyset_{1a}$, i_{in} is sampled to the *coarse* memory transistor. Then, in phase $\emptyset_{1b}$, the current stored in the coarse transistor is compared to the actual value of i_{in} and the difference is stored in the *fine* memory transistor. During the output phase $\emptyset_2$, the value in the fine memory is subtracted from the coarse sample, and thus the CFT error from the coarse sampling is removed. Unfortunately, the fine memory sampling switch also adds a CFT error. This error is has a *weak signal dependency* since it is a function of the magnitude of the coarse CFT. As a first order approximation it can be regarded as constant. It can be made signal independent to a higher order by using the S^nI cell. It is essentially an extension of S^2I to an *n*-step algorithm, using $(n-1)$ fine memories. As seen from Fig. 2.19 b, the S^nI cell proposed in [68] also uses all NMOS memory transistors, which is likely to increase speed.

As mentioned above, the dummy switch technique has been combined with S^2I and S^nI in order to remove the constant CFT residue [57-58]. A number of other non-ideal effects were addressed as the *enhanced* and *seamless* S^2I memory cells were proposed [69-70]. Another variation on the two-step theme is found in [28] where both the fine and coarse memory transistors are NMOS. A very simple form of "algorithmic" CFT cancellation can be used to cancel the constant CFT error. Because of the inherent signal inversion in memory cells, two cascaded memory cells will produce errors that tend to cancel each other. Unfortunately, this is only true for the constant CFT component. The signal dependent errors are approximately doubled, as shown by the author in [32]. Yet, a reduction in error *magnitude* is achieved. This approach was used in [71] where a double

NMOS - PMOS memory cell was used. A similar effect is seen in fully-differential SI circuits, which are treated below.

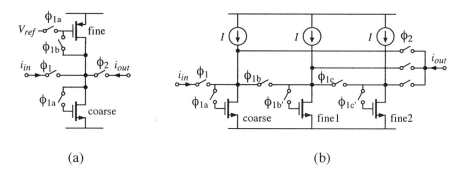

Fig. 2.19 Algorithmic CFT cancellation. (a) S^2I memory cell [67]. (b) S^nI cell [68].

2.3.4 Fully-differential, feedforward and feedback techniques

Fully-differential SI circuits have inherent cancellation of the signal independent part of the CFT, since it will appear as a common-mode term, which is suppressed by the *common-mode rejection ratio* (CMRR). A number of fully-differential SI memory cells have been presented, e. g. [2, 72-74] The relative amount of signal-dependent CFT remains more or less unchanged since it appears in the differential signal [32]. *Common-mode feedback* (CMFB) [75-76] or *common-mode feedforward* (CMFF) [77] techniques can be used to further suppress common-mode errors.

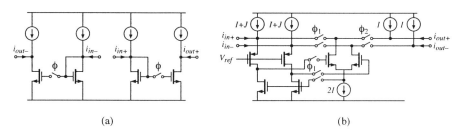

Fig. 2.20 Fully-differential SI circuits. (a) Simple fully-differential realizations [2]. (b) Fully-differential realization in [72].

An *error feedback* technique, remotely related to the S^2I technique was used in [78] to reduce CFT errors. Similar to S^2I, a coarse sampling is followed by a fine-tuning phase. In [78], the fine-tuning is performed by a *negative feedback loop* that adjusts the gate voltage on the hold transistor until the held current equals the input current. *Oversampling SI memory cells* using feedback [79] and feedforward [80] architectures have also been proposed.

2.3.5 Adaptive techniques

The clock signal is often the target for *adaptive* CFT cancellation schemes. As shown previously, the amount of CFT is a function of clock slope and clock voltage levels. In [18, 26], an adaptive scheme was used to make a clock signal with reduced swing track the sampling switch threshold voltage, thereby achieving near-constant CFT. A similar approach was used in [81] where only the clock "low" voltage is regulated. Dummy switch cancellation is a simple technique, with the drawback that it is very sensitive to clock edge slope and timing. An adaptive scheme can be used to optimize the effectiveness and robustness of the dummy switch technique [82-83]. By adjusting the clock skew between the real switch and the dummy, optimum cancellation of constant CFT is achieved.

2.3.6 Zero-voltage switching techniques

Signal-dependent CFT can be eliminated in *switched-capacitor* circuits by the use of *zero-voltage switching*. By keeping the sampling switch at a constant potential, a "virtual ground", there will be no signal dependent CFT errors. A zero-voltage switching SI memory cell was presented in [84] and is shown in Fig. 2.21 a. Negative feedback keeps the input node at a fixed potential and therefore the sampling switch S1 always injects the same amount of charge. If extended to a fully-differential circuit, the constant CFT is reduced by the common-mode rejection, as demonstrated in [85].

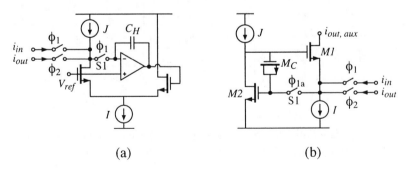

Fig. 2.21 Zero-voltage switching SI circuits. (a) Circuit proposed in [84]. (b) Circuit proposed in [86].

An entirely new memory cell, shown in Fig. 2.21 b, was proposed in [86]. Its current input and output is on the *source* node of the memory transistor. A feedback loop keeps the input voltage at the (constant) value of V_{GS2} that is required for M2 to sink the bias current J. Thus, the sampling switch S1 is always operated at the same potential, and the CFT error becomes signal independent. Other zero-voltage switching SI memory cells have been proposed in [87] where an autotuning approach was suggested, and in [88] where class-AB techniques were used.

2.3.7 CFT compensation techniques used in this book

First generation SI circuits were used throughout this book. In the author's early work regarding wave SI filters [89], the *dummy-circuit cancellation* technique was used. In a later publication [49], the *complete cancellation* technique proposed by the author was used, as described in chapter 8. The latter was also used for the Δ-Σ modulator implementation in [50, 90] and chapter 9 where its efficiency is experimentally verified. Both wide-band A/D converter implementations in [91] and [92] (see chapters 10 and 11) utilizes the *signal dependent cancellation* technique proposed by Fiez/Song. A *fully-differential* version with *common-mode feedforward* according to [77] was used.

2.4 Current sample-and-hold evolution

The *current sample-and-hold* (*CSH*) or *memory cell* (MC) is the core building block of any SI circuit. Most of the performance of an SI circuit is also determined by the quality of the CSH. At present time, SI circuits have been a research topic for about twelve years, and a great number of improvements on the basic SI memory cell have been proposed. The evolution of the SI memory cell as seen from publications in the open literature is reviewed in this section. Included references were carefully chosen with emphasis on who was *first* to publish the idea and it has been the author's intention to cover *all* novel SI memory cell realizations. Obviously, there may be a few missing or badly chosen publications. The feasibility or performance of each individual realization is *not* evaluated, but its most distinguishing features are mentioned briefly in the following evolution history.

1988

[3]   "Current-copier". Later known as "2nd generation MC". Conductance ratio improvements: Cascode and active-mirror type MC. CFT compensation: Negative feedback loop sampling.

1989

[2]   1st generation MC. Basic fully-differential realizations.

1990

[5]   BiCMOS OTA-based MC.
[18] 2nd generation regulated cascode MC. Adaptive clock voltage scheme gives a near-constant CFT.
[25] Replica-based constant CFT cancellation. Dummy switch.
[93] Conveyor MC. Resembles the current-conveyor.
[94] Algorithmic MC. Achieves CFT cancellation through an iterative sampling scheme.

1991

[95] 2nd generation class AB MC.
[61] … identical to:
[26] Replica-based signal-dependent CFT cancellation for 1st generation MC.

1992

[82] Dummy-switch with adaptive clock-delay. Minimize constant CFT.
[72] Fully-differential 2nd generation MC using one differential pair.
[96] Voltage delay lines using SI MC.
[97-98] 1st generation GaAs MC, including fully-differential realization.
[73] Fully-differential 2nd generation MC using two differential pairs (N & P).
[99] 2nd generation BiCMOS MC.
[22] Transimpedance "active" MC improving conductance ratio and speed.

1993

[67] S^2I – two step CFT cancellation.
[4]   CMOS OTA-based MC.
[48] 1st generation MC with replica-based complete CFT cancellation and arbitrary positive/negative gain proposed by the author.
[100] 2nd generation regulated cascode, class AB MC.

1994

- [75] 2nd generation low-voltage, fully-differential, class AB MC with common-mode feedback (CMFB).
- [101] 2nd generation and S^2I GaAs MC.
- [81] Adaptive clock voltage. Clock LOW tracks input voltage in order to remove signal-dependent CFT.
- [68] S^nI – multi step CFT cancellation.
- [84] Zero-voltage switching for SI circuits. CFT becomes signal-independent.
- [102] Class AB S^nI.
- [52] CFT attenuation through Miller-capacitance.
- [79] Oversampling MC with feedback structure.
- [78] CFT compensation by negative error feedback. The sampled current is fine-tuned by comparing it to the input signal.
- [103] High-speed BiCMOS fully-differential MC.
- [71] Double MC (PMOS & NMOS). Constant CFT term is essentially canceled after passing through a S/H pair. This was also shown by the author in [29].
- [77] Common-mode feedforward (CMFF). Cancel all common-mode errors, e.g. constant CFT.
- [65] 1st generation MC with "double" replica-based complete CFT cancellation.

1995

- [58] S^nI combined with replica-based constant CFT cancellation.
- [53] CFT cancellation by fully-differential Miller-capacitance.
- [64] 2nd generation "replica-style" total CFT cancellation through combination of switch channel sizes.
- [76] Fully-differential MC with CMFB.
- [74] Ultra-low voltage, high-speed MC. Internally fully-differential – single-ended in/out.
- [56] 2nd generation combined with replica-based constant CFT cancellation.
- [104] Novel voltage memory based on SI MC.

1996

- [105] Novel BiCMOS fully-differential MC.
- [69] Enhanced S^2I. Compensated for voltage drop on output switch. Neutralization of drain-gate capacitance in fully differential realization. Reset of "fine" memory improves settling.
- [106] Reduction of signal-dependent CFT through optimization of charge-injection in CMOS switch.
- [86] New MC architecture. In/out on memory transistor source node. Virtual-ground input gives zero-voltage switching. No signal-dependent CFT.

1997

- [57] S^2I combined with replica-based constant CFT cancellation.
- [70] S^3I – Seamless S^2I.

1998

[28] Modified S^{2}I approach where the coarse and fine memory transistors are both NMOS, and operated mostly concurrently.
[54] New MC using common-gate amp and dummy-switch. Common-gate amp yields a constant CFT which is canceled by the dummy switch.
[80] Oversampling MC with feedforward structure.

1999

[88] Zero-voltage switching class AB MC.
[66] Error neutralization by cross-coupled neutralization cap in fully-differential MC.
[107] Floating gate class AB MC.

REFERENCES

[1] P. E. Allen, and D. R. Holberg, *CMOS Analog Circuit Design*, Holt, Rinehart, and Winston, Inc., New York, 1987.

[2] J. B. Hughes, N. C. Bird, and I. C. Macbeth, "Switched Currents – A New Technique for Analog Sampled-Data Signal Processing", *Proc. of Int. Symp. Circuits And Systems (ISCAS)*, Portland, Oregon, pp. 1584-1587, May 1989, IEEE.

[3] S. J. Daubert, D. Vallancourt, and Y. P. Tsividis, "Current Copier Cells", *Electron. Lett.*, Vol. 24, No. 25, pp. 1560-1562, Dec. 1988.

[4] T. Inoue, Q. Pan, F. Ueno, and Y. Ohuchi, "Design and Analysis of OTA Switched Current Mirrors", IEICE Trans. Fundamentals, Vol. E76-A, No. 6, pp. 940-946, June 1993.

[5] D. Robertson, P. Real, and C. Mangelsdorf, "A Wideband 10-bit, 20Msps Pipelined ADC using Current-Mode Signals", *Proceedings of IEEE Solid-State Circ. Conf.*, San Francisco, California, pp. 206-207, Feb. 1990, IEEE.

[6] G. Bogason, *Switched Current Circuits - Design, Optimization and Application*, Ph.D. dissertation, Technical University of Denmark, Sept. 1995.

[7] D. G. Nairn, and C. A. T. Salama, "Algorithmic Analog/Digital Convertor Based on Current Mirrors", *Electron. Lett.*, Vol. 24, No. 8, pp. 471-472, Apr. 1988.

[8] H. Träff, "A Novel Approach to High Speed CMOS Current Comparators", *Electron. Lett.*, Vol. 28, No. 3, pp. 310-312, Jan. 1992.

[9] G. Di Cataldo, and G. Palumbo, "New CMOS Current Schmitt Triggers", *Proc. of ISCAS 92*, San Diego, California, pp. 1292-1295, May 1992, IEEE.

[10] A. T. K. Tang, and C. Toumazou, "High-Performance CMOS Current Comparator", *Electron. Lett.*, Vol. 30, No. 1, pp. 5-6, Jan. 1994.

[11] G. Palmisano, and G. Palumbo, "Offset Compensation Technique for CMOS Current Comparators", *Electron. Lett.*, Vol. 30, No. 11, pp. 852-854, May 1994.

[12] A. Worapishet, J. B. Hughes, and C. Toumazou, "Enhanced Switched-Current Comparator", *Electron. Lett.*, Vol. 35, No. 10, pp. 767-768, May 1999.

[13] A. Worapishet, J. B. Hughes, and C. Toumazou, "Error Neutralised Switched-Current Comparator", *Proc. of 1999 Int. Symp. Circuits and Systems*, Orlando, Florida, Vol. 2, pp. 464-467, May 1999, IEEE.

[14] A. Rodríguez-Vázquez, R. Domínguez-Castro, F. Medeiro, and M. Delgado-Restituto, "High-Resolution CMOS Current Comparators: Design and Applications to Current-Mode Function Generation", *Analog Integrated Circuits and Signal Processing*, Vol. 7, No. 2, pp. 149-165, Mar. 1995.

[15] N. C. Battersby, *Switched-Current Techniques for Analogue Sampled-Data Signal Processing*, Ph. D. dissertation, Imperial College, University of London, Feb. 1993

[16] D. G. Nairn, and C. A. T. Salama, "Algorithmic analogue-to-digital convertors using current-mode techniques", *IEE Proceedings*, Vol. 137, *Pt G*, No. 2, pp. 163-168, Apr. 1990.

[17] E. Säckinger, and W. Guggenbühl, "High-Swing, High-Impedance MOS Cascode Circuit", *IEEE J. Solid-State Circ.*, Vol. 25, No. 1, pp. 289-298, Feb. 1990.

[18] C. Toumazou, J. B. Hughes, and D. M. Patullo, "Regulated Cascode Switched-Current Memory Cell", *Electron. Lett.*, Vol. 26, No. 5, pp. 303-305, Mar. 1990.

[19] P. J. Crawley, and G. W. Roberts, "High-Swing MOS Current Mirror with Arbitrarily High Output Resistance", *Electron. Lett.*, Vol. 28, No. 4, pp. 361-363, Feb. 1992.

[20] T. Loeliger, and W. Guggenbühl, "Cascode Configurations for Switched-Current Copiers", *Analog Integrated Circuits and Signal Processing*, Vol. 19, No. 2, pp. 115-127, May 1999.

[21] D. G. Nairn, and C. A. T. Salama, "High-Resolution, Current-Mode A/D Convertors Using Active Current Mirrors", *Electron. Lett.*, Vol. 24, No. 21, pp. 1331-1332, Oct. 1988.

[22] D. G. Nairn, "Amplifiers for High-Speed Current-Mode Sample-and-Hold Circuits", *Proc. of Int. Symp. Circuits And Systems (ISCAS)*, San Diego, California, pp. 2045-2048, May 1992, IEEE.

[23] E. Bruun, "Dynamic Range of Low-Voltage Cascode Current Mirrors", *Proc. of Int. Symp. Circuits And Systems (ISCAS)*, Seattle, Washington, pp. 1328-1331, May 1995, IEEE.

[24] R. T. Baird, T. S. Fiez, D. J. Allstot, "Speed and Accuracy Considerations in Switched-Current Circuits", *Proc. of Int. Symp. Circuits And Systems (ISCAS)*, Singapore, pp. 1809-1812, June 1991, IEEE.

[25] H. C. Yang, T. S. Fiez, and D. J. Allstot, "Current-Feedthrough Effects and Cancellation Techniques in Switched-Current Circuits", *Proc. of Int. Symp. Circuits And Systems (ISCAS)*, pp. 3186-3188, May 1990, New Orleans, IEEE.

[26] T. S. Fiez, D. J. Allstot, G. Liang, and P. Lao, "Signal-Dependent Clock-Feedthrough Cancellation in Switched-Current circuits", *Proc. of China 1991 Int. Conf. Circuits And Systems*, Shenzhen, China, pp. 785-788, June 1991, IEEE.

[27] M. Song, Y. Lee, and W. Kim, "A Clock Feedthrough Reduction Circuit for Switched-Current Systems", *IEEE J. Solid-State Circ.*, Vol. 28, No. 2, Feb. 1993, pp. 133-137.

[28] M. Helfenstein, and G. S. Moschytz, "Improved Two-Step Clock-Feedthrough Compensation Technique for Switched-Current Circuits", *IEEE Trans. on CAS-II*, Vol. 45, No. 6, pp. 739-743, June 1998.

[29] D. Macq, and P. Jespers, "Charge Injection in Current-Copier Cells", *Electron. Lett.*, Vol. 29, No. 9, pp. 780-781, Apr. 1993.

[30] G. Wegmann, E. A. Vittoz, and F. Rahali, "Charge Injection in Analog MOS Switches", *IEEE J. Solid-State Circ.*, Vol. SC-22, No. 6, Dec. 1987, pp. 1091-1097.

[31] J.-H. Shieh, M. Patil, and B. J. Sheu "Measurement and Analysis of Charge Injection in MOS Analog Switches", *IEEE J. Solid-State Circ.*, Vol. SC-22, No. 2, Apr. 1987, pp. 277-281.

[32] B. Jonsson, *Applications of the Switched-Current Technique*, Licentiate Thesis No. 458, Linköping University, Sweden, Oct. 1994.
[33] J. R. Burns, "Large-Signal Transit-Time Effects in the MOS-Transistor", *RCA Rev.*, vol. 15, pp. 14-35, Mar. 1969
[34] C. D. Motchenbacher, and J. A. Conelly: *Low-Noise Electronic System Design*, Wiley, 1993.
[35] K. R. Laker, and W. M. C. Sansen: *Design of Analog Circuits and Systems*, McGraw-Hill, 1994.
[36] S. J. Daubert, and D. Vallancourt, "Noise Analysis of Current Copier Circuits", *Proc. of Int. Symp. Circuits And Systems (ISCAS)*, New Orleans, Louisiana, pp. 307-310, May 1990, IEEE.
[37] D. G. Nairn, and C. A. T. Salama, "A Ratio-Independent Algorithmic Analog-to-Digital Converter Combining Current Mode and Dynamic Techniques", *IEEE Trans. Circuits Syst.*, Vol. 37, No. 3, pp. 319-325, Mar. 1990.
[38] C. Toumazou, F. J. Lidgey, D. G. Haigh, (Eds.) *Analogue IC design: the current-mode approach*, IEE Circuits and Systems series 2, 1990.
[39] P. Shah, and C. Toumazou, "A Theoretical Basis for Very Wide Dynamic Range Switched-Current Analogue Signal Processing", *Analog Integrated Circuits and Signal Processing*, Vol. 7, No. 3, pp. 201-213, May 1995.
[40] I. H. Jørgensen, and G. Bogason, "Noise Analysis of Switched-Current Circuits", *Proc. of Int. Symp. Circuits And Systems (ISCAS)*, Monterey, CA., Vol. 1, pp. 108-111, May 1998, IEEE.
[41] J.-S. Wang, R. Huang, and C.-L. Wey, "Synthesis of Optimal Current-Copiers for Low-Power/Low-Voltage Switched-Current Circuits", *Proceedings of IEEE Midwest Symp. Circ. Syst.*, Notre Dame, Indiana, pp. 220-223, Aug. 1998, IEEE.
[42] I. H. H. Jørgensen, and G. Bogason, "Noise Analysis of Switched-Current Circuits", *Analog Integrated Circuits and Signal Processing*, Vol. 18, No. 1, pp. 69-71, Jan. 1999.
[43] J. B. Hughes, *Analogue Techniques for Very Large Scale Integrated Circuits*, Ph. D. dissertation, University of Southampton, UK, March 1992
[44] B. E. Jonsson, "Design of Power Supply Lines in High-Performance SI and Current-Mode Circuits", *Proc. of 15th NORCHIP Conf.*, Tallinn, Estonia, pp. 245-250, Nov. 1997, IEEE.
[45] R. v. d. Plassche, *Integrated Analog-to-Digital and Digital-to-Analog Converters*, pp. 6-9, Kluwer Academic Publishers, 1994.
[46] B. Jonsson, S. Signell, H. Stenström, and N. Tan, "Distortion in Sampling", *Proc. of Int. Symp. Circuits And Systems (ISCAS)*, Hong Kong, Vol. 1, pp. 445-448, June 1997, IEEE.
[47] B. E. Jonsson, "Sampling Jitter in High-Speed SI Circuits", *Proc. of Int. Symp. Circuits And Systems (ISCAS)*, Monterey, California, Vol. 1, pp. 524-526, May 1998, IEEE.
[48] B. Jonsson, and S. Eriksson, "New Clock-Feedthrough Compensation Scheme for Switched-Current Circuits", *Electron. Lett.*, Vol. 29, No. 16, pp. 1446-1447, Aug. 1993.
[49] B. Jonsson, and S. Eriksson, "A Low Voltage Wave SI Filter Implementation Using Improved Delay Elements", *Proc. of Int. Symp. Circuits And Systems (ISCAS)*, London, UK, pp. 5.305-5.308, May 1994, IEEE.

[50] B. Jonsson, and N. Tan, "Clock-Feedthrough Compensated First-Generation SI Circuits and Systems", *Analog Integrated Circuits and Signal Processing*, Vol. 12, No. 4, pp. 201-210, Apr. 1997.

[51] C. Wang, M. Omair-Ahmad, and M. N. S. Swamy, "Design of a Transistor-Mismatch-Insensitive Switched-Current Memory Cell", *Proc. of 1999 Int. Symp. Circuits and Systems*, Orlando, Florida, Vol. 2, pp. 105-108, May 1999, IEEE.

[52] W. Guggenbühl, J. Di, and J. Goette, "Switched-Current Memory Circuits for High-Precision", *IEEE J. Solid-State Circ.*, Vol. 29, No. 9, pp. 1108-1116, Sept. 1994.

[53] C.-Y. Wu, C.-C. Cheng, and J.-J. Cho, "Precise CMOS Current Sample/Hold Circuits Using Differential Clock Feedthrough Attenuation Techniques", *IEEE J. Solid-State Circ.*, Vol. 30, No. 1, pp. 76-80, Jan. 1995.

[54] K. Leelavattananon, J. B. Hughes, and C. Toumazou, "Very Low Charge Injection Switched-Current Memory Cell", *Proc. of Int. Symp. Circuits And Systems (ISCAS)*, Monterey, CA., Vol. 1, pp. 531-534, May 1998, IEEE.

[55] D. M. W. Leenaerts, G. R. M. Hamm, M. J. Rutten, and G. G. Persoon, "High-Performance Switched-Current Memory Cell", *Proceedings of ECCTD 97*, Budapest, Hungary, Vol. 1, pp. 234-239, Sept. 1997.

[56] C. K. Tse, and M. H. L. Chow, "A New Clock-Feedthrough Cancellation Method for Second-Generation Switched-Current Circuits" *Proc. of Int. Symp. Circuits And Systems (ISCAS)*, Seattle, Washington, pp. 2104-2107, May 1995, IEEE.

[57] H.-W. Cha, S. Ogawa, and G. Watanabe, "A Clock-Feedthrough Compensated Switched-Current Memory Cell", *IEICE Trans. Fund. El. Comm. And Comp. Sci.*, Vol. E80-A, No. 6, pp. 1069-1071, June 1997.

[58] X. Zeng, C. K. Tse, and P. S. Tang, "A New Scheme for Complete Cancellation of Charge Injection Distortion in Second Generation Switched-Current Circuits", *Proceedings of 1995 IEEE Region 10 International Conference on Microelectronics and VLSI*, Hong Kong, pp. 127-130, Nov. 1995.

[59] X. Zeng, C. K. Tse, and P. S. Tang, "Total Charge Injection Cancellation Scheme for High-Precision Second-Generation Switched-Current Circuits", *Proceedings of ISCAS 96*, Atlanta, Georgia, Vol. 1, pp. 413-416, May 1996, IEEE.

[60] X. Zeng, C. K. Tse, and P. S. Tang, "Complete Clock-Feedthrough Cancellation in Switched-Current Circuits by Combining the Replica Technique and the n-step Principle", *Analog Integrated Circuits and Signal Processing*, Vol. 12, No. 2, pp. 145-152, Feb. 1997.

[61] M. Song, Y. Lee, and W. Kim, "A New Design Methodology of Second Order Switched-Current Filter", *Proc. of Int. Symp. Circuits And Systems (ISCAS)*, Singapore, pp. 1797-1800, June 1991, IEEE.

[62] B.-M. Min, and S.-W. Kim, "New Clock-Feedthrough Compensation Scheme for Switched-Current Circuits", *IEEE Trans. on CAS-II*, Vol. 45, No. 11, pp. 1508-1511, Nov. 1998.

[63] T. Takagiwa, A. Hyogo, and K. Sekine, "A Signal-Dependent Clock Feedthrough Cancellation Technique for Switched-Current Circuits", *Electronics and Communications in Japan Part 2 (Electronics)*, Vol. 80, No. 9, pp. 26-34, Sept. 1997.

[64] M. Helfenstein, and G. S. Moschytz, "Clockfeedthrough Compensation Technique for Switched-Current Circuits", *IEEE Trans. on CAS-II*, Vol. 42, No. 3, pp. 229-231, Mar. 1995.

[65] H.-K. Yang, and E. I. El-Masry, "Clock Feedthrough Analysis and Cancellation in Current Sample/Hold Circuits", *IEE Proc. Pt. G.*, Vol. 141, No. 6, pp. 510-516, Dec. 1994.

[66] J. B. Hughes, and K. W. Moulding, "Error Neutralisation in Switched Current Memory Cells", *Proc. of 1999 Int. Symp. Circuits and Systems*, Orlando, Florida, Vol. 2, pp. 460-463, May 1999, IEEE.

[67] J. B. Hughes, and K. W. Moulding, "S^2I: A Two-Step Approach to Switched-Currents", *Proc. of Int. Symp. Circuits And Systems (ISCAS)*, Chicago, Illinois, pp. 1235-1238, May 1993, IEEE.

[68] C. Toumazou, and S. Xiao, "n-step charge injection cancellation scheme for very accurate switched current circuits", *Electron. Lett.*, Vol. 30, No. 9, pp. 680-681, Apr. 1994.

[69] J. B. Hughes, and K. W. Moulding, "Enhanced S^2I Switched-Current Cells", *Proceedings of ISCAS 96*, Atlanta, Georgia, Vol. 1, pp. 187-190, May 1996, IEEE.

[70] J. B. Hughes, and K. W. Moulding, "S^3I: The Seamless S^2I Switched-Current Cell", *Proc. of Int. Symp. Circuits And Systems (ISCAS)*, Hong Kong, Vol. 1, pp. 113-116, June 1997, IEEE.

[71] D. M. W. Leenaerts, A. J. Leeuwenburgh, and G. G. Persoon, "A High-Performance SI Memory Cell", *IEEE J. Solid-State Circ.*, Vol. 29, No. 11, pp. 1404-1407, Nov. 1994.

[72] J. B. Hughes, and K. W. Moulding, "Switched-Current Video Signal Processing", *Proc. of Custom Int. Circ. Conf*, Boston, Massachusetts, pp. 24.4/1-4, May 1992.

[73] U. Gatti, and F. Maloberti, "Fully Differential Switched-Current Building Blocks", *Proc. of Int. Symp. Circuits And Systems (ISCAS)*, San Diego, California, pp. 1400-1403, May 1992, IEEE.

[74] Y. Sugimoto, "A 1.6V 10-bit 20MHz Current-Mode Sample and Hold Circuit", *Proc. of Int. Symp. Circuits And Systems (ISCAS)*, Seattle, Washington, pp. 1332-1335, May 1995, IEEE.

[75] H. Träff, and S. Eriksson, "Novel Pseudo-Class AB Fully Differential 3V Switched-Current System Cells", *Electron. Lett.*, Vol. 30, No. 7, pp. 536-537, Mar. 1994.

[76] C.-Y. Wu, C.-C. Chen, and J.-J. Cho, "A CMOS Transistor-Only 8-b 4.5Ms/s Pipelined Analog-to-Digital Converter using Fully-Differential Current-Mode Circuit Techniques", *IEEE J. Solid-State Circuits.*, Vol. 30, No. 5, pp. 522-532, May. 1995.

[77] N. Tan, and S. Eriksson, "Low-Voltage Fully Differential Class-AB SI Circuits with Common-Mode Feedforward", *Electron. Lett.*, Vol. 30, No. 25, pp. 2090-2091, Dec. 1994.

[78] B. Pain, and E. R. Fossum, "A Current Memory Cell with Switch Feedthrough Reduction by Error Feedback", *IEEE J. Solid-State Circ.*, Vol. 29, No. 10, pp. 1288-1290, Oct. 1994.

[79] I. Mehr, and T. Sculley, "A 16-Bit Current Sample/Hold Circuit Using a Digital Process", *Proc. of Int. Symp. Circuits And Systems (ISCAS)*, London, UK, pp. 5.417-5.420, May 1994, IEEE.

[80] R. Huang, and C.-L. Wey, "A High-Accuracy CMOS Oversampling Switched-Current Sample/Hold (S/H) Circuit using Feedforward Approach", *IEEE Trans. on CAS-II*, Vol. 45, No. 3, pp. 395-399, June 1998.

[81] J. Simek, M. Bojcun, and D. Valehrachova, "Class AB Switched-Current Building Blocks for Sampled-Data Filters", *Elektrotechnicky Casopis*, Vol. 45, No. 4, pp. 124-128, 1994.

[82] S. Espejo, A. Rodriguez-Vazquez, R. Dominguez-Castro, and J. L. Huertas, "An Adaptive Scheme for Feedthrough Cancellation in Switched-Current Techniques", *Proceedings of IEEE Midwest Symp. Circ. Syst.*, Washington, DC, Vol. 2, pp. 1292-1295, Aug. 1992, IEEE.

[83] S. Espejo, A. Rodriguez-Vazquez, R. Dominguez-Castro, and J. L. Huertas, "A Modified Dummy-Switch Technique for Tunable Feedthrough Cancellation in Switched-Current Circuits", *Proceedings of ESSCIRC 93*, Sevilla, Spain, pp. 270-273, Sept. 1993.

[84] D. G. Nairn, "Zero-Voltage Switching in Switched-Current Circuits", *Proc. of Int. Symp. Circuits And Systems (ISCAS)*, London, UK, pp. 5.289-5.292, May 1994, IEEE.

[85] J. M. Martins, and V. F. Dias, "Very Low-Distortion Fully Differential Switched-Current Memory Cell", *IEEE Trans. on CAS-II*, Vol. 46, No. 5, pp. 640-643, May 1999.

[86] P. Shah, and C. Toumazou, "A New High Speed Low Distortion Switched-Current Cell", *Proceedings of ISCAS 96*, Atlanta, Georgia, Vol. 1, pp. 421-424, May 1996, IEEE.

[87] H. Ishii, S. Takagi, and N. Fujii, "Switched Current Circuit using Fixed Gate Potential and Automatic Tuning Circuit", *Electrical Engineering in Japan*, Vol. 126, No. 3, pp. 21-29, Feb. 1999.

[88] A. Worapishet, J. B. Hughes, and C. Toumazou, "Class AB Technique for High Performance Switched-Current Memory Cells", *Proc. of 1999 Int. Symp. Circuits and Systems*, Orlando, Florida, Vol. 2, pp. 456-459, May 1999, IEEE.

[89] B. Jonsson, and S. Eriksson, "Current-Mode N-port Adaptors for Wave SI Filters", *Electron. Lett.*, Vol. 29, No. 10, pp. 925-926, May 1993.

[90] N. Tan, B. Jonsson, and S. Eriksson, "3.3V 11bit Delta-Sigma Modulator using First-Generation SI Circuits", *Electron. Lett.*, Vol. 30, No. 22, pp. 1819-1821, Oct. 1994.

[91] B. E. Jonsson, and H. Tenhunen, "A 3V Switched-Current Pipelined Analog-to-Digital Converter in a 5V CMOS process", *Proc. of 1999 Int. Symp. Circuits and Systems*, Orlando, Florida, Vol. 2, pp. 351-354, May 1999, IEEE.

[92] B. E. Jonsson, and H. Tenhunen, "A Dual 3-V 32-MS/s CMOS Switched-Current ADC for Telecommunication Applications", *Proc. of 1999 Int. Symp. Circuits and Systems*, Orlando, Florida, Vol. 2, pp. 343-346, May 1999, IEEE.

[93] J. B. Hughes, I. C. Macbeth, and D. M. Patullo, "Switched-Current System Cells", *Proc. of Int. Symp. Circuits And Systems (ISCAS)*, New Orleans, Louisiana, pp. 303-306, May 1990, IEEE.

[94] C. Toumazou, N. C. Battersby, C. Maglaras, "High-Performance Algorithmic Switched-Current Memory Cell", *Electron. Lett.*, Vol. 26, No. 19, pp. 1593-1595, Sept. 1990.

[95] N. C. Battersby, and C. Toumazou, "Class AB Switched-Current Memory for Analogue Sampled-Data Systems", *Electron. Lett.*, Vol. 27, No. 10, pp. 873-875, May 1991.

[96] N. Tan, and S. Eriksson, "High Performance Voltage Delay Lines Using Switched-Current Memory Cells", *Electron. Lett.*, Vol. 28, No. 3, pp. 228-229, Jan. 1992.

[97] C. Toumazou, and N. C. Battersby, "High Speed GaAs Switched Current Techniques for Analogue Sampled-Data Signal Processing", *Electron. Lett.*, Vol. 28, No. 7, pp. 689-690, Mar. 1992.

[98] C. Toumazou, N. Battersby, and M. Puwani, "GaAs Switched-Current Techniques for Front-End Analogue Signal Processing Applications", *Proceedings of IEEE Midwest Symp. Circ. Syst.*, Washington, DC, Vol. 1, pp. 44-47, Aug. 1992, IEEE.

[99] B. H. Leung, "BiCMOS Based Self-Calibrated Current Cells for High-Precision Digital to Analog Converters", *Proc. of Int. Symp. Circuits And Systems (ISCAS)*, San Diego, California, pp. 2348-2351, May 1992, IEEE.

[100] H. Träff, and S. Eriksson, "Class A and AB Compact Switched-Current Memory Circuits", *Electron. Lett.*, Vol. 29, No. 16, pp. 1454-1455, Aug. 1993.
[101] S. Xiao, and C. Toumazou, "Second Generation Single and Two-Step GaAs Switched-Current Cells", *Electron. Lett.*, Vol. 30, No. 9, pp. 681-683, Apr. 1994.
[102] C. Toumazou, and G. Saether, "Switched-current circuits and systems", *Proc. of Int. Symp. Circuits And Systems (ISCAS)*, London, UK, May 1994, Vol. Tutorials, pp. 459-486, IEEE.
[103] T. Reimann, F. Krummenacher, and M. Declercq, "High Speed BiCMOS Current Mode Differential Track-and-Hold Circuit", *Electron. Lett.*, Vol. 30, No. 21, pp. 1730-1732, Oct. 1994.
[104] D. Lachartre, "Switched-Current Low-Power Fast Adressable Analogue Memory", *Electron. Lett.*, Vol. 31, No. 21, pp. 1808-1809, Oct. 1995.
[105] X. Hu, "A Switched-Current Sample-and-Hold Circuit", *Proceedings of ISCAS 96*, Atlanta, Georgia, Vol. 1, pp. 191-194, May 1996, IEEE.
[106] S. Lindfors, K. Halonen, "A Novel Clock Feed-Through Distortion Cancellation Method for SI Circuits", *Proceedings of ISCAS 96*, Atlanta, Georgia, Vol. 1, pp. 61-64, May 1996, IEEE.
[107] I. Mucha, "Ultra Low Voltage Class AB Switched Current Memory Cells based on Floating Gate Transistors", *Analog Integrated Circuits and Signal Processing*, Vol. 20, No. 1, pp. 43-62, July 1999.

Chapter 3

Switched-Current Systems

Switched-current circuits have found applications in sampled-data signal processing and data-conversion. This chapter serves as to give an overview of what has been done in the fields of sampled-data filters, Nyquist A/D-converters and oversampling A/D-converters. The intention is to give references for further reading, and to define a context for the implementations and theoretical work described in the remainder of the book.

3.1 Sampled-data filters

SI sampled-data filters were proposed in [1-2], and a lot more work has followed. The author's contribution to this field is the *N-port adaptor* for wave SI filters [3] and associated work [4]. This is covered in chapter 8 of this book. In order to get a broader view, FIR and ladder filters are also included in this review. Some work on design methodology and analysis is mentioned, as well as a few other types of filter. Where illustrations are given, early or basic realizations are used as examples, while later and more improved realizations are referred to in the text. Cascode transistors have been omitted in all circuit schematics for simplicity.

3.1.1 FIR-filters

FIR filters are characterized by their *finite impulse response*. They have the advantage that, for a given FIR filter response, there is always a stable circuit realization, and that it is possible to realize FIR filters with linear phase [5]. A traditional FIR filter realization is shown in Fig. 3.1 a. A delay line is used to store the M previous samples of the input signal which are tapped, weighted, and combined to form the output signal. A switched-current circuit realizing the delay and the tap scaling coefficient is shown in Fig. 3.1 b. Switched-current FIR filters of this type are found in [6-8]. A straight-forward voltage-mode implementation of this filter would require one OP-amp per delay stage. That would lead to an unacceptably large area and power dissipation as M is increased. The advantage of a switched-current implementation is that the much simpler circuit shown in Fig. 3.1 b is sufficient. A *digitally programmable* SI FIR filter with a slightly different approach is found in [9]. Instead of propagating the signal though the delay line, the digitally stored coefficient values are propagated while each sample remains in its memory for M clock cycles. Class-AB current-copier cells were used as delay cells, and digitally controlled current-division circuits [10] were used to implement the filter coefficients. Another digitally programmable SI filter is described in [11], where a circular FIR structure was used.

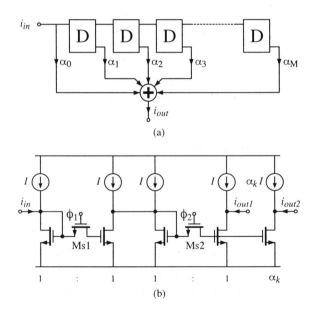

Fig. 3.1 (a) FIR filter schematics. (b) First-generation SI realization of FIR filter "slice" including one full delay and a weighted tap.

3.1.2 IIR-filters

With IIR filters there is more freedom to shape the filter response since both poles and zeros can be placed freely as long as stability is guaranteed.

Examples of IIR filters are discrete-time wave filters and integrator-based filters, such as ladder filters, and biquadratic filters.

3.1.2.1 Integrator-based filters

Ladder filters simulate the currents and voltages in passive LC ladder filters [12], and the low sensitivity to component tolerances is maintained in the transformation. The continuous-time s variable is mapped onto the discrete-time z by some transformation, e. g. the bilinear transformation [5]

$$s \rightarrow \frac{2}{T} \frac{1-z^{-1}}{1+z^{+1}} \qquad (3.1)$$

An example of the filters that result is shown in Fig. 3.2 a [7]. The basic building block is the bilinear integrator in Fig. 3.2 b. Filter coefficients are implemented by a direct scaling of the integrator output currents, as shown in Fig. 3.2 b. A large number of SI ladder filters have been reported in the literature. Both first-generation [7-8, 12] and second-generation [13-14] SI memory cells have been used. Class-AB integrator realizations were shown in [15] and [16], and the use of GaAs SI memory cells for high-speed implementations was demonstrated in [17-18]. A high-speed CMOS implementation is reported in [19], where a double-sampling S^2I approach [20] was used to reach an 80 MS/s sampling rate. The measured THD is ~ 48 dB with a 1 MHz input signal. Switched-current ladder *band-pass* filters are treated in [21].

Second-order *biquadratic sections* can be cascaded in order to realize higher-order filter functions. Each section realizes a generic second-order transfer function

$$H(z) = \frac{A + Bz^{-1} + Cz^{-2}}{1 + Dz^{-1} + Ez^{-2}} \qquad (3.2)$$

This approach has been used in several SI filter publications, e.g., [1, 22-24]. *Digitally programmable* second-order sections without transmission zeros were proposed in [25], where binary weighted current-mirrors were used to control center frequency, selectivity (Q), and input gain. In [26], *autotuning* was used to accurately control an OTA-based SI realization of the same type of second-order section.

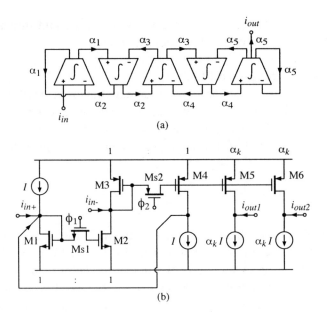

Fig. 3.2 (a) Ladder filter schematics. (b) First-generation SI realization of integrator used in [7] with non-inverting and inverting inputs and dual weighted outputs.

3.1.2.2 Wave SI filters

Wave digital filters (WDF) have been proposed and treated in detail by Fettweis [27-28]. They simulate the wave propagation and reflections in a microwave filter. A discrete-time wave filter is built from delay elements and *adaptors*. The delay elements simulate the propagation delay in transmission lines, and the adaptors simulate the wave transmission and reflections that occur at impedance discontinuities. Wave SI filters (WSIF) were proposed in [29] where filters using two-port adaptors were realized. In order to realize WSIF with transmission zeros, such as elliptical filters, three-port adaptors are needed. The author therefore proposed circuit realizations for general N-port serial and parallel adaptors in [3]. Another group independently developed programmable three-port adaptors which were demonstrated shortly thereafter in [30]. More detail on that work is found in [31], and the author's contributions are described further in [3] and [4] (see chapter 8). A WSIF sensitivity analysis is performed in [31]. Using a different transformation of the reference filter leads to slightly different circuit realizations in [32] which illustrates a variation on the WSIF theme.

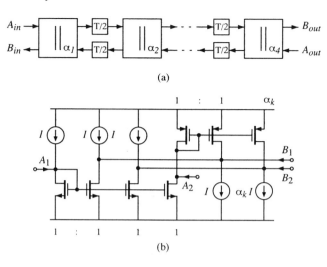

Fig. 3.3 (a) Wave SI filter schematics. (b) Current-mode realization of two-port adaptor used in [29].

3.1.3 Miscellaneous filters

A number of other SI filter realizations have been proposed. Among them are the *direct-form* SI filters in [33] which were extended to be digitally programmable in [34]. *Multirate* filters for FIR and IIR interpolation and decimation were proposed in [35]. A decimator implementation for HDTV applications is presented in [36]. Switched-current *median filters* were demonstrated in [37].

3.1.4 Design methodology and analysis

Tools and methodology for SC filter design have been developed over a long time. For SI circuits it has been necessary to develop the corresponding tools and methods. One way to quickly overcome the lack of specialized SI design tools is to do an SC design and then map it onto an SI realization. This was demonstrated in [38] where the signal flowgraph (SFG) of an SC filter was transposed to give an SFG suitable for SI circuit implementation. The direct mapping of ladder filter SFGs onto SI circuits without passing through an intermediate SC filter was used in [13]. Ultimately, the filter response can be extracted from SPICE level simulations. There is however a need for methods to analyze the filter response at a higher level of abstraction. Such analysis is discussed in [39-43]. Related details are treated in [44]. An "inspection" method to derive the SFG directly from the schematic representation of an SI filter is proposed and described in [45-47].

Extensive theoretical work has also been presented by Ng and Sewell, covering *matrix decomposition methods* for wideband filter design [48], and *N-path* and *pseudo-N-path techniques* for narrow-band filters [49]. Further, they have treated *SI decimator* and *interpolator* filters [50], and presented techniques for reducing the sensitivity of *high order* SI filters [51].

3.1.5 Influence of SI circuit imperfections

In the design of SI filters, it is essential to understand how circuit imperfections influence the final performance. A brief summary of the severity of different errors are given below:
- Constant offset errors at various nodes of the filter are usually harmless since in a linear filter it will only create a DC offset at the output. In many cases this is no serious problem as the offset is easily removed.
- Linear gain errors alter the filter coefficients, and that is a much more serious problem since it results in a pole-zero displacement and distortion of the frequency-domain transfer function. Most of the SI circuit imperfections, such as CFT, mismatch and conductance ratio errors, give a linear error component in addition to the signal-dependent and constant offsets. Accurate transfer function implementation is therefore highly dependent on the overall quality of the SI building blocks. A tolerance of 0.1 to 1% can be expected with careful design and layout.
- Nonlinear errors always contribute to harmonic distortion, and the severity of these errors is very much dependent on the design specification.

3.2 A/D converters

A/D converter design has been a part of the SI field of research from the start [52-54]. The author's contributions to this field are the *SI delta-sigma modulator* in [55] (see chapter 9), the *wideband SI ADC core* in [56] (see chapter 10), and the *high-speed parallel time-interleaved SI ADC* in [57] (see chapter 11). The distinction between *switched-current* and *current-mode* ADCs is not always obvious. Moreover, some implementations use mixed-mode signals, i.e., current *and* voltage. Due to the scope of this book, the text below will focus on implementations that are truly current-mode, and particularly those that incorporate sampling as a part of the conversion process. The review of SI ADCs is subdivided into Nyquist and oversampling A/D converters. Just as in the previous "*Sampled-data filters*" section, early or basic realizations are used for illustrations, while later and more improved realizations are referred to in the text.

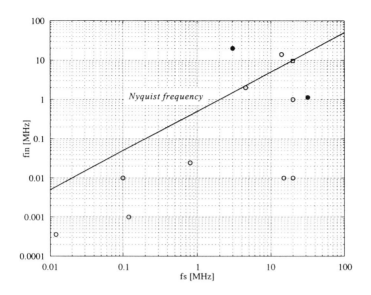

Fig. 3.4 Measured f_{in} and f_s for Nyquist SI and current-mode A/D converters. Legend: (o) CMOS, (□) BiCMOS, and (●) ADCs included in this book.

3.2.1 Nyquist A/D converters

Nyquist ADCs ideally have an input signal bandwidth, BW_{in}, that is one half of the sampling frequency, f_s. Various implementations of such converters are reviewed in the following subsections. In general, it is found that few of the reported SI ADC implementations were characterized over their full input bandwidth. Input frequencies well below 1 MHz have been used in most cases, regardless of sampling rate. Therefore their reported performance is more comparable to *narrow-band* oversampling ADCs. This is illustrated in Fig. 3.4, which shows the f_s and f_{in} at which performance was reported for different implementations. Several of the early publications do not report which input frequency was used for the measurements. The designs that report Nyquist performance are [58-60] and the ADC core found in chapter 10 of this book [56]. Designs with $BW_{in} \geq 1$ MHz are [58-62] and the Nyquist ADCs in this book [56-57]. All of these implementations are further compared in chapter 10.

3.2.1.1 Pipelined and cyclic ADCs

The early work by Nairn et. al. [63] laid the foundation for SI A/D converters, although it is itself not *SI* but rather *current-mode*. A time-continuous cascade of bitcells was used to implement the A/D conversion algorithm described by Eqs. 3.3 and 3.4.

$$i_{k+1} = 2i_k - b_k I_{ref}$$
$$b_k = \begin{cases} 1 & i_k > I_{ref} \\ 0 & \text{otherwise} \end{cases} \quad (3.3)$$

$$i_0 = i_{in}$$
$$i_{in} \in \{0 \ldots 2I_{ref}\} \quad (3.4)$$

It was followed by a *true* SI implementation where second-generation current memory cells were used to realize a mismatch-free gain of two [53]. Its circuit realization is shown in Fig. 3.5. The input current i_{in} is first sampled by M1, then by M2. The sum ($2i_{in}$) is then loaded into M3 and compared to I_{ref}. If the bit decision $b_{out} = 1$, I_{ref} is subtracted from the output current of this step. In the **cyclic** implementation in [53], the output current is again sampled by M1 and M2, and the process is repeated. This gives a very compact implementation at the cost of reduced conversion speed. The 10-b cyclic implementation in [53] was measured to have an INL of 0.92 LSB at 25 kS/s. Similar cyclic ADC implementations are found in [64-66]. The highest resolution, INL = +/- 1 LSB @ 14-b and 5.7 kS/s, is reported in [64]. The highest sampling rate, 120 kS/s, is reported in [65], while the design in [66] has the smallest chip area, 0.014 mm².

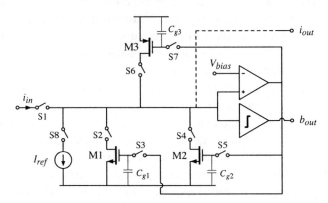

Fig. 3.5 Circuit realization of the cyclic A/D converter in [53].

In order to have a *1-b/stage* **pipelined** implementation of the same algorithm, i_{out} can be fed to the following stage as indicated in Fig. 3.5. This is essentially what was done in [54]. A similar 2 V pipeline ADC with bi-directional input currents was proposed in [67]. The highest reported *effective number-of-bits* (ENOB) for any CMOS switched-current ADC with

$f_s \geq 1$ MHz is found in [60], where this algorithm is also used. Its measured ENOB is **8.18** at 4.5 MS/s with $f_{in} = 200$ kHz.

Exact implementation of the interstage gain of two is essential for ADC linearity, but may limit conversion speed. A current-mirror with ratio 1:2 is slower than a unity-gain mirror and its accuracy is limited by mismatch. The dynamic multiplication of two shown in Fig 3.5 requires one extra clock phase for the sampling and is therefore unnecessarily slow. By successively scaling I_{ref} with 1/2 and having an interstage gain of one, this problem is solved. This approach was used for the designs in [68-69]. Very high sample rates were achieved – 20 MS/s in [68] and 14 MS/s with $f_{in} \sim 14$ MHz in [69]. The conversion algorithm for bi-directional input signals is described by Eq. 3.5.

$$i_0 = i_{in}$$
$$i_{k+1} = i_k + (-1)^{b_k} \frac{I_{ref}}{2^{k+1}} \quad (3.5)$$
$$b_k = \begin{cases} 1 & i_k > 0 \\ 0 & \text{otherwise} \end{cases}$$

Another reference-dividing approach was used in [52] where a pipelined successive approximation algorithm was implemented.

The *RSD*, or *1.5-b/stage,* algorithm has become popular due to its significantly relaxed offset requirements [70]. Redundant information that is extracted in each "bitcell" is used for digital correction. The first SI implementation by Macq et. al. is found in [71], where an INL ≤ 0.8 LSB @ 10-b was measured at 550 kS/s. A few other designs has followed, e.g., [62, 72-74]. The design in [62] has a measured S/(N+THD) ratio of ~ 34.2 dB for a 1 MHz input signal sampled at 20 MHz. This is equivalent to 5.65 effective bits. In [74] 10.5 effective bits were measured for a 22.4 kHz input signal sampled at 800 kHz. An RSD architecture was also used by the author for the ADCs described in chapters 10 and 11 [56-57]. A peak resolution of **7.43** effective bits at 3 MS/s, and a 20 MHz input bandwidth was achieved.

Switched-current *multibit/stage pipeline ADCs* with digital correction have been presented in [58, 75]. OTA-based BiCMOS memory cells were used in [58] to achieve a measured ENOB of approximately 7.7 for a 10 MHz input signal sampled at 20 MHz. Even though the power dissipation is as high as 1 W, this represents the best performance reported for any wideband SI ADC to this date.

Parallel time-interleaved architectures can be used to increase sample rate. A problem with such structures is to take care of the channel offset and gain errors. The ADC in [59] was used in a 4x-parallel structure [61], and measurements showed that distortion due to channel mismatch errors were less than -50 dB at 20 MS/s. Regular SNR and THD measures were not reported. A parallel ADC designed by the author [57] achieved 32 MS/s sample rate by using eight parallel core ADCs [56]. This is the highest sampling rate for which measured results have been reported for *any* switched-current ADC. Measured SFDR is 50 dB after channel calibration. The design is presented in chapter 11.

An *ultra-low voltage* cyclic SI ADC is found in [76]. The measured INL is 1.4 LSB @ 10-b when sampled at 12 kS/s. An error correction algorithm similar to what is often used in multibit pipelined ADCs is used for this 1-b/cycle implementation operating at 1.5 V.

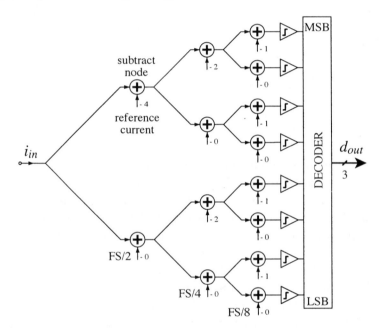

Fig. 3.6 A 3-b simplified version of the current-tree flash A/D converter architecture in [78].

3.2.1.2 Flash ADCs

In a flash converter, all decision levels are evaluated simultaneously by approximately 2^N comparators. While the comparators in a voltage-mode flash ADC can share a single copy of the input signal, it is generally required to generate one input copy per comparator in a current-mode flash. This can be done by *splitting* the input current into fractions [77] or by generating *multiple copies* thereof [78]. The current-tree structure of [78] is shown in Fig. 3.6. Reference currents are not inserted at the comparators but rather subtracted from the signal copies "along the way" in a distributed fashion. One way to reduce the required number of comparators is the *two-step flash* [79]. A *digitally calibrated* two-step flash was proposed in [80].

3.2.1.3 Miscellaneous ADCs

A very respectable dynamic performance was reported in [81]. Measured peak THD and ENOB are -79 dB and 9.1 bits at 200 kS/s for a 10-b current-mode ADC. A successive approximation architecture with a MOSFET-only *current division* [10] R-2R ladder DAC was used. The circuit realization of the 10-b DAC, which is the heart of the design, is shown in Fig. 3.7.

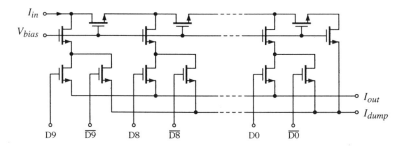

Fig. 3.7 MOST-only current division R-2R DAC used for the ADC in [81].

Finally, it can be mentioned that many *folding and interpolating* ADCs use a current-mode signal in the actual folding/interpolating circuitry, while the rest of the ADC is essentially voltage-mode, e.g., [82]. This type of ADC is not considered in this book.

3.2.2 Oversampling Δ-Σ A/D converters

Oversampling Δ-Σ A/D converters have gained a lot of interest lately. By forming a feedback loop around the quantizer, the quantization noise can be spectrally shaped and placed out of the signal band. The one-bit Δ-Σ converter is probably the most popular architecture due to the inherent linearity of 1-b A/D and D/A converters. A lot of work has been done on Δ-Σ A/D conversion in general, i.e., [83-84], and on switched-current delta-sigma converters in particular, e.g., [85]. It seems that nearly all SI Δ-Σ converters use 1-b quantization, and most often a second-order LP architecture. This architecture was also used by the author in [55, 86]. Second-generation or S^2I memory cells are used almost exclusively, except in the work by the author where first-generation SI memory cells were used. Theoretically, second-generation memory cells should realize a better unity-gain sample-and-hold, and therefore automatically be a better candidate for realization of integrators. It has been shown, however, that without a good

CFT compensation, the linear component of the CFT error becomes equivalent to a gain error [86]. In a comparison between eight different SI Δ-Σ modulator implementations [87], it was found that the one using first-generation memory cells proposed by the author [55] also had the highest resolution (11-b). Nevertheless, since then higher performance has been reported for S^2I-based Δ-Σ modulators in [88], which demonstrated the highest performance reported for any SI implementation, 14-b linearity and SNR > 80 dB.

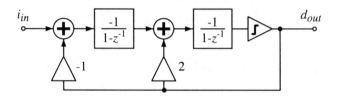

Fig. 3.8 Second order delta-sigma modulator architecture used in [90].

The first reported switched-current Δ-Σ modulator (DSM) is found in [89]. OTA-based SI memory cells were used to implement a second-order DSM for voice-band applications with 13-b resolution. A second-order DSM using second-generation SI memory cells were proposed in [90] together with simulation results. Measured results from early SI DSM implementations are found in [91] and [55]. The latter includes work by the author. Other measured second-order implementations are found in [92-94].

Third [88] and *fourth order* [95] SI DSM have been proposed, as well as higher-order *bandpass* modulators [96-97]. A technique for *self-calibration* of integrator gain was demonstrated in [98], and built-in self-test (BIST) in [99]. How to increase dynamic range effectively by *power scaling* was shown in [100], and a complex *mixed analog-digital* DSM architecture was proposed in [101]. The authors claim lower distortion through reduced current swing in the integrators. The analog signal is split into a digital part and an analog residue which are processed separately. Simulations are used to confirm the benefits of the architecture. Finally, a detailed description of the optimization and design of an SI DSM is given in [102], where a number of published designs are also compared.

3.2.3   Influence of SI circuit imperfections

Most A/D conversion architectures can be described by some algorithm including one or more of the operations: comparison, scaling, addition and sampling. In general, A/D converter architectures are sensitive to the accuracy of all of these operations. All kind of errors may cause the

comparator to make the wrong decision, and even if a particular error does not immediately cause a wrong decision, errors accumulate throughout the conversion, possibly causing bit errors in later stages.
- Constant offset errors are generally a problem in ADC implementations. Because an ADC is built from highly non-linear circuits, such as comparators, even small offsets can create large errors. In general, offset errors result in all sorts of output errors: offset, gain errors and harmonic distortion. Therefore offset insensitive ADC architectures are usually a good choice for SI implementations.
- Linear gain errors alter the interstage gain in pipelined ADCs. In oversampling delta-sigma ADCs, the noise transfer function and possibly the stability of the modulator loop is changed. Linear gain errors in general cause harmonic distortion and an ADC gain error.
- Nonlinear errors almost inevitably result in harmonic distortion. The exception is that, if the comparator input signal is distorted by a non-linearity without DC offset, it will not add any errors. Non-linear errors in Δ-Σ modulators can also increase the in-band noise through intermodulation between large amplitude out-of-band noise components [88].

REFERENCES

[1] J. B. Hughes, N. C. Bird, and I. C. Macbeth, "Switched Currents – A New Technique for Analog Sampled-Data Signal Processing", *Proc. of Int. Symp. Circuits And Systems (ISCAS)*, Portland, Oregon, pp. 1584-1587, May 1989, IEEE.

[2] D. Vallancourt, and Y. P. Tsividis, "Sampled-Current Circuits", *Proc. of Int. Symp. Circuits And Systems (ISCAS)*, Portland, Oregon, pp. 1592-1595, May 1989, IEEE.

[3] B. Jonsson, and S. Eriksson, "Current-Mode N-port Adaptors for Wave SI Filters", *Electron. Lett.*, Vol. 29, No. 10, pp. 925-926, May 1993.

[4] B. Jonsson, and S. Eriksson, "A Low Voltage Wave SI Filter Implementation Using Improved Delay Elements", *Proc. of Int. Symp. Circuits And Systems (ISCAS)*, London, UK, pp. 5.305-5.308, May 1994, IEEE.

[5] L. Wanhammar, *DSP Integrated Circuits*, Academic Press Series in Engineering, 1999.

[6] G. Liang, and D. J. Allstot, "FIR Filtering Using CMOS Switched-Current Techniques", *Proc. of Int. Symp. Circuits And Systems (ISCAS)*, New Orleans, Louisiana, pp. 2291-2293, May 1990, IEEE.

[7] D. J. Allstot, T. S. Fiez, and G. Liang, "Design Considerations for CMOS Switched-Current Filters", *Proc. of Custom Int. Circ. Conf*, Boston, Massachusetts, pp. 8.1/1-4, May 1990.

[8] T. S. Fiez, G. Liang, and D. J. Allstot, "Switched-Current Circuit Design Issues", *IEEE J. Solid-State Circuits*, Vol. 26, No. 3, pp. 192-202, Mar. 1991.

[9] Y. L. Cheung, and A. Buchwald, "A Sampled-Data Switched-Current Analog 16-Tap FIR Filter with Digitally Programmable Coefficients in 0.8μm CMOS", *Proc. of Int. Solid-State Circ. Conf.*, San Francisco, California, pp. 54-55, 429, Feb. 1997, IEEE.

[10] K. Bult, and G. Geelen, "An inherently linear and compact MOST-only current division technique", *IEEE J. Solid State Circ.*, Vol. 27, pp. 1730-1735, Dec. 1992.

[11] F. A. Farag, C. Galup-Montoro, and M. C. Schneider, "A Programmable Low-Voltage Switched-Current FIR Filter", *Proc. of 1999 Int. Symp. Circuits and Systems*, Orlando, Florida, Vol. 2, pp. 472-475, May 1999, IEEE.

[12] T. S. Fiez, and D. J. Allstot, "CMOS Switched-Current Ladder Filters", *IEEE J. Solid-State Circuits*, Vol. 25, No. 6, pp. 1360-1367, Dec. 1990.

[13] A. C. M. de Queiroz, and P. R. M. Pinheiro, "Exact Design of Switched-Current Ladder Filters", *Proc. of Int. Symp. Circuits And Systems (ISCAS)*, San Diego, California, pp. 855-858, May 1992, IEEE.

[14] N. C. Battersby, and C. Toumazou, "A 5th Order Bilinear Elliptic Switched-Current Filter", *Proc. of Custom Int. Circ. Conf*, San Diego, California, pp. 6.3/1-4, May 1993.

[15] N. C. Battersby, and C. Toumazou, "Class AB Switched-Current Memory for Analogue Sampled-Data Systems", *Electron. Lett.*, Vol. 27, No. 10, pp. 873-875, May 1991.

[16] H. Träff, and S. Eriksson, "Class A and AB Compact Switched-Current Memory Circuits", *Electron. Lett.*, Vol. 29, No. 16, pp. 1454-1455, Aug. 1993.

[17] N. C. Battersby, and C. Toumazou, "Towards High Frequency Switched-Current Filters in CMOS and GaAs", *Proc. of Int. Symp. Circuits And Systems (ISCAS)*, Chicago, Illinois, pp. 1239-1242, May 1993, IEEE.

[18] S. Xiao, and C. Toumazou, "Second Generation Single and Two-Step GaAs Switched-Current Cells", *Electron. Lett.*, Vol. 30, No. 9, pp. 681-683, Apr. 1994.

[19] J. B. Hughes, and K. W. Moulding, "An 8MHz, 80Ms/s Switched-Current Filter", *Proc. of IEEE Int. Solid-State Circ. Conf.*, San Francisco, California, pp. 60-61, Feb. 1994, IEEE.

[20] J. B. Hughes, and K. W. Moulding, "A Switched-Current Double Sampling Bilinear Z-Transform Filter Technique", *Proc. of Int. Symp. Circuits And Systems (ISCAS)*, London, UK, pp. 5.293-5.296, May 1994, IEEE.

[21] A. C. M. de Queiroz, and P, M. Pinheiro, "Switched-Current Ladder Band-Pass Filters", *Proc. of Int. Symp. Circuits And Systems (ISCAS)*, London, UK, pp. 5.309-5.312, May 1994, IEEE.

[22] T. S. Fiez, B. Lee, and D. J. Allstot, "CMOS Switched-Current Biquadratic Filters", *Proc. of Int. Symp. Circuits And Systems (ISCAS)*, New Orleans, Louisiana, pp. 2300-2303, May 1990, IEEE.

[23] J. B. Hughes, I. C. Macbeth, and D. M. Patullo, "Switched-Current System Cells", *Proc. of Int. Symp. Circuits And Systems (ISCAS)*, New Orleans, Louisiana, pp. 303-306, May 1990, IEEE.

[24] J. B. Hughes, I. C. Macbeth, and D. M. Patullo, "Second Generation Switched-Current Signal Processing", *Proc. of Int. Symp. Circuits And Systems (ISCAS)*, New Orleans, Louisiana, pp. 2805-2808, May 1990, IEEE.

[25] A. G. Begisi, T. S. Fiez, and D. J. Allstot, "Digitally-Programmable Switched-Current Filters", *Proc. of Int. Symp. Circuits And Systems (ISCAS)*, New Orleans, Louisiana, pp. 3178-3181, May 1990, IEEE.

[26] Y. Ohuchi, T. Inoue, and H. Fujino, "A Design of Switched-Current Auto-Tuning Filter and its Analysis", *IEICE Trans. Fundamentals*, Vol. E78-A, No. 10, pp. 1350-1354, Oct. 1995.

[27] A. Fettweis, "Digital Filter Structures Related to Classical Filter Networks", *Arch. Elektr. Übertragungstech.*, Vol. 25, No. 2, pp. 79-89, Feb. 1971.

[28] A. Fettweis, "Wave Digital Filters: Theory and Practice", *IEEE Proceedings*, Vol. 74, No. 2, pp. 270-327, Feb. 1986.

[29] A. Rueda, A. Yúfera, J. L. Huertas, "Wave analogue filters using switched-current techniques", *Electron. Lett.*, 1991, Vol. 27, No. 16, pp. 1482-1483, Aug. 1991.

[30] A. Yúfera, A. Rueda, J. L. Huertas, "A Methodology for Programmable Switched-Current Filters Design", *Proc. of European Conf. Circuit Theory and Design*, Davos, Switzerland, pp. 317-322, Aug. 1993, Elsevier.

[31] A. Yúfera, A. Rueda, and J. L. Huertas, "Programmable Switched-Current Wave Analog Filters", *IEEE J. Solid-State Circ.*, Vol. 29, No. 8, pp. 927-935, Aug. 1994.

[32] J. D. Lancaster, B. M. Al-Hashimi, and M. Moniri, "Efficient SI Wave Elliptic Filters based on Direct and Inverse Bruton Transformations", *IEE Proc. Pt. G.*, Vol. 146, No. 5, pp. 235-241, Oct. 1999.

[33] S. Wang, and M. O. Ahmad, "A new Design of Switched-Current IIR Filters", *Proceedings of 1995 Canadian Conf. on Electrical and Computer Eng.*, Montreal, Que., Canada, Vol. 1, pp. 461-464, Sept. 1995.

[34] M. O. Ahmad, and S. Wang, "A Novel Fully Programmable Switched-Current IIR Filter", *Proc. of Custom Int. Circ. Conf*, Santa Clara, California, pp. 12.4/1-4, May 1997.

[35] W. Ping, and J. E. Franca, "Switched-Current Multirate Filtering", *Proc. of Int. Symp. Circuits And Systems (ISCAS)*, London, UK, pp. 5.321-5.324, May 1994, IEEE.

[36] M. Helfenstein, J. E. Franca, and G. S. Moschytz, "Design Techniques for HDTV Switched-Current Decimators", *Proceedings of ISCAS 96*, Atlanta, Georgia, Vol. 1, pp. 195-198, May 1996, IEEE.

[37] C. K. Tse, and K. C. Chun, "Design of a Switched-Current Median Filter", *IEEE Trans. on CAS-II*, Vol. 42, No. 5, pp. 356-359, May 1995.

[38] G. W. Roberts, and A. S. Sedra, "Synthesizing Switched-Current Filters By Transposing the SFG of Switched-Capacitor Filter Circuits", *IEEE Trans. Circuits Syst.*, Vol. 38, No. 3, pp. 337-340, Mar. 1991.

[39] A. C. M. de Queiroz, P. R. M. Pinheiro, and L. P. Calôba, "Systematic Nodal Analysis of Switched-Current Filters", *Proc. of Int. Symp. Circuits And Systems (ISCAS)*, Singapore, pp. 1801-1804, June 1991, IEEE.

[40] A. C. M. de Queiroz, P. R. M. Pinheiro, and L. P. Calôba, "Nodal Analysis of Switched-Current Filters", *IEEE Trans. Circuits Syst.–II*, Vol. 40, No. 1, pp. 10-18, Jan. 1993.

[41] E. M. Schneider, and T. S. Fiez, "Simulation of Switched-Current Systems", *Proc. of Int. Symp. Circuits And Systems (ISCAS)*, Chicago, Illinois, pp. 1420-1423, May 1993, IEEE.

[42] J. A. Barby, "Switched-Current Filter Models for Frequency Analysis in the Continuous-Time Domain", *Proc. of Int. Symp. Circuits And Systems (ISCAS)*, Chicago, Illinois, pp. 1427-1430, May 1993, IEEE.

[43] Z. O. Shang, and J. I. Sewell, "Development of Efficient Switched Network and Mixed-Mode Simulators", *IEE Proc. Pt. G.*, Vol. 145, No. 1, pp. 24-34, Feb. 1998.

[44] A. C. M. de Queiroz, and P. R. M. Pinheiro, "Switching Sequence Effects in Switched-Current Filters", *Proc. of Int. Symp. Circuits And Systems (ISCAS)*, Chicago, Illinois, pp. 982-985, May 1993, IEEE.

[45] M. Helfenstein, A. Muralt, and G. S. Moschytz, "Direct Analysis of Multiphase Switched-Current Networks using Signal-Flowgraphs", *Proc. of Int. Symp. Circuits And Systems (ISCAS)*, Seattle, Washington, pp. 1476-1479, May 1995, IEEE.

[46] M. Helfenstein, *Analysis and Design of Switched-Current Networks*, Series in Microelectronics, Vol. 70, Hartung-Gorre Verlag Konstantz, 1997.

[47] M. Helfenstein, A. Muralt, and G. S. Moschytz, "Direct Analysis and Synthesis of Multiphase Switched-Current Networks using Signal-Flow Graphs", *Int. Journal of Circuit Theory and Applications*, Vol. 26, No. 3, pp. 253-280, May-June 1998.

[48] A. E. J. Ng, and J. I. Sewell, "Ladder Decompositions for Wideband SI Filter Applications", *IEE Proc. Pt. G.*, Vol. 145, No. 5, pp. 306-313, Oct. 1998.

[49] A. E. J. Ng, and J. I. Sewell, "N-path and Pseudo-N-Path Cells for Switched-Current Signal Processing", *IEEE Trans. on CAS-II*, Vol. 46, No. 9, pp. 1148-1160, Sept. 1999.

[50] A. E. J. Ng, and J. I. Sewell, "Ladder Derived Switched-Current Decimators and Interpolators", *IEEE Trans. on CAS-II*, Vol. 46, No. 9, pp. 1161-1170, Sept. 1999.

[51] A. E. J. Ng, and J. I. Sewell, "Feasible Designs for High Order Switched-Current Filters", *IEE Proc. Pt. G.*, Vol. 145, No. 5, pp. 297-305, Oct. 1998.

[52] P. Deval, G. Wegmann, and J. Robert, "CMOS Pipelined A/D Convertor Using Current Divider", *Electron. Lett.*, Vol. 25, No. 20, pp. 1341-1342, Sept. 1989.

[53] D. G. Nairn, and C. A. T. Salama, "Ratio-Independent Current-Mode Algorithmic Analog-To-Digital Converters", *Proceedings of ISCAS 89*, Portland, Oregon, pp. 250-253, May 1989, IEEE.

[54] J. Robert, P. Deval, and G. Wegmann, "Novel CMOS Pipelined A/D Convertor Architecture Using Current Mirrors", *Electron. Lett.*, Vol. 25, No. 11, pp. 691-692, May 1989.

[55] N. Tan, B. Jonsson, and S. Eriksson, "3.3V 11bit Delta-Sigma Modulator using First-Generation SI Circuits", *Electron. Lett.*, Vol. 30, No. 22, pp. 1819-1821, Oct. 1994.

[56] B. E. Jonsson, and H. Tenhunen, "A Low-Voltage, 10-b Switched-Current ADC with 20 MHz Input Bandwidth", *Electron. Lett.*, Vol. 34, No. 20, pp. 1904-1905, Oct. 1998.

[57] B. E. Jonsson, and H. Tenhunen, "A Low-Voltage 32MS/s Parallel Pipelined Switched-Current ADC", *Electron. Lett.*, Vol. 34, No. 20, pp. 1906-1907, Oct. 1998.

[58] P. Real, D. H. Robertson, and C. W. Mangelsdorf, "A Wideband 10-b 20-Ms/s Pipelined ADC Using Current-Mode Signals", *IEEE J. Solid-State Circuits.*, Vol. 26, No. 8, pp. 1103-1109, Aug. 1991.

[59] M. Bracey, W. Redman-White, J. Richardson, and J. B. Hughes, "A Full Nyquist 15 MS/s 8-bit Differential Switched-Current A/D Converter", *Proceedings of ESSCIRC 95*, Lille, France, pp. 146-149, Sept. 1995.

[60] C.-Y. Wu, C.-C. Chen, and J.-J. Cho, "A CMOS Transistor-Only 8-b 4.5Ms/s Pipelined Analog-to-Digital Converter using Fully-Differential Current-Mode Circuit Techniques", *IEEE J. Solid-State Circuits.*, Vol. 30, No. 5, pp. 522-532, May. 1995.

[61] M. Bracey, W. Redman-White, J. B. Hughes, and J. Richardson, "A 70 MS/s 8-bit Differential Switched-Current CMOS A/D Converter Using Parallel Interleaved Pipelines", *Proceedings of 1995 IEEE Region 10 International Conference on Microelectronics and VLSI*, Hong Kong, pp. 143-146, Nov. 1995.

[62] Y. Sugimoto, and T. Iida, "A Low-Voltage, High-Speed and Low-Power Full Current-Mode Video-rate CMOS A/D Converter", *Proceedings of ESSCIRC 97*, Southampton, UK, pp. 392-395, Sept. 1997.

[63] D. G. Nairn, and C. A. T. Salama, "Algorithmic Analog/Digital Convertor Based on Current Mirrors", *Electron. Lett.*, Vol. 24, No. 8, pp. 471-472, Apr. 1988.

[64] P. Deval, J. Robert, and M. J. Declercq, "A 14-bit CMOS A/D Converter Based on Dynamic Current Memories", *Proc. of Custom Int. Circ. Conf*, San Diego, California, pp. 24.2/1-4, May 1991.

[65] S.-W. Kim, and S.-W. Kim, "Current-Mode Cyclic ADC for Low-Power and High-Speed Applications", *Electron. Lett.*, Vol. 27, No. 10, pp. 818-820, May 1991.

[66] M. Kondo, H. Onodera, and K. Tamaru, "A Current-Mode Cyclic A/D Converter with Submicron Processes", *IEICE Trans. Fund. El. Comm. And Comp. Sci.*, Vol. E80-A, No. 2, pp. 360-364, Feb. 1997.

[67] L. Zhang, T. Sculley, and T. Fiez, "A 12 Bit, 2V Current-Mode Pipelined A/D Converter using a Digital CMOS Process", *Proceedings of ISCAS 94*, London, UK, pp. 5.369-5.372, May 1994, IEEE.

[68] M. Gustavsson, *Analog Interfaces in a Digital CMOS Process*, Licentiate Thesis No. 662, Linköping University, Sweden, Dec. 1997.

[69] M. Bracey, W. Redman-White, J. Richardson, and J. B. Hughes, "A Full Nyquist 15 MS/s 8-b Differential Switched-Current A/D Converter", *IEEE J. Solid-State Circuits.*, Vol. 31, No. 7, pp. 945-951, July 1996.

[70] B. Ginetti, P. G. A. Jespers, and A. Vandemeulebroecke, "A CMOS 13-b Cyclic RSD A/D Converter", *IEEE J. Solid State Circ.*, Vol. 27, No. 7, pp. 957-964, July 1994.

[71] D. Macq, and P. G. A. Jespers, "A 10-bit Pipelined Switched-Current A/D Converter", *IEEE J. Solid-State Circuits.*, Vol. 29, No. 8, pp. 967-971, Aug. 1994.

[72] R. C. Hui, and H. C. Luong, "A CMOS Current-Mode Pipeline ADC using Zero-Voltage Sampling Technique", *Proc. of Int. Symp. Circuits And Systems (ISCAS)*, Monterey, CA., Vol. 1, pp. 9-12, May 1998, IEEE.

[73] J.-S. Wang, and C.-L. Wey, "A 12-b, 100ns/b, 1.9mW Switched-Current Cyclic A/D Converter", *Proc. of Int. Symp. Circuits And Systems (ISCAS)*, Monterey, CA., Vol. 1, pp. 416-419, May 1998, IEEE.

[74] J.-S. Wang, and C.-L. Wey, "A 12-bit 100ns/bit 1.9-mW CMOS Switched-Current Cyclic A/D Converter", *IEEE Trans. on CAS-II*, Vol. 46, No. 5, pp. 507-516, May 1999.

[75] J. P. A. Carreira, C. Dupuy, and J. E. Franca, "A Compact Three-Step Pipelined CMOS Current-Mode A/D Converter", *Proceedings of ISCAS 97*, Hong Kong, Vol. 1, pp. 465-468, June 1997, IEEE.

[76] C.-C. Cheng, and C.-Y. Wu, "Design Techniques for 1.5-V Low-Power CMOS Current-Mode Cyclic Analog-to-Digital Converters", *IEEE Trans. Circuits Syst.–II*, Vol. 45, No. 1, pp. 28-40, Jan. 1998.

[77] W. R. Krenik, R. K. Hester, R. D. DeGroat, "Current-Mode Flash A/D Conversion Based On Current-Splitting Techniques", *Proceedings of ISCAS 92*, San Diego, California, pp. 585-588, May 1992, IEEE.

[78] H. Hasegawa, M. Yotsuyanagi, M. Yamaguchi, and K. Sone, "A 1.5 V Video-Speed Current-Mode Current Tree A/D Converter", *Proceedings of IEEE Symp. VLSI Circ.*, Honolulu, Hawaii, pp. 17-18, June 1994, IEEE.

[79] A. Cable, and R. Harjani, "A 6-Bit 50MHz Current-Subtracting Two Step Flash Converter", *Proceedings of ISCAS 94*, London, UK, pp. 5.465-5.468, May 1994, IEEE.

[80] J. P. Oliveira, J. Vital, and J. E. Franca, "A Digitally Calibrated Current-Mode Two-Step Flash A/D Converter", *Proceedings of ISCAS 96*, Atlanta, Georgia, Vol. 1, pp. 199-202, May 1996, IEEE.

[81] C. Hammerschmied, and Q. Huang, "Design and Implementation of an Untrimmed MOSFET-Only, 10-Bit A/D Converter with –79-dB THD", *IEEE J. Solid-State Circuits.*, Vol. 33, No. 8, pp. 1148-1157, Aug. 1998.

[82] M. P. Flynn, and D. J. Allstot, "CMOS Folding A/D Converters with Current-Mode Interpolation", *IEEE J. Solid-State Circuits.*, Vol. 31, No. 9, pp. 1248-1257, Sept. 1996.

[83] J. C. Candy, and G. C. Temes (Eds.): *Oversampling Delta-Sigma Data Converters: Theory, Design and Simulation*, IEEE Press, 1992.

[84] S. R. Norsworthy, R. Schreier, and G. C. Temes (Eds.): *Delta-Sigma Data Converters: Theory, Design and Simulation*, IEEE Press, 1997.

[85] N. Tan, *Switched-Current Design and Implementation of Oversampling A/D Converters*, Kluwer, 1997.

[86] B. Jonsson, and N. Tan, "Clock-Feedthrough Compensated First-Generation SI Circuits and Systems", *Analog Integrated Circuits and Signal Processing*, Vol. 12, No. 4, pp. 201-210, Apr. 1997.

[87] N. Tan, "Switched-Current Delta-Sigma A/D Converters", *Analog Integrated Circuits and Signal Processing*, Vol. 9, No. 1, pp. 7-24, Jan. 1996.

[88] N. Moeneclaey, and A. Kaiser, "Design Techniques for High-Resolution Current-Mode Sigma-Delta Modulators", *IEEE J. Solid-State Circ.*, Vol. 32, No. 7, pp. 953-958, July 1997.

[89] S. J. Daubert, and D. Vallancourt, "A Transistor-Only Current-Mode $\Sigma\Delta$ Modulator", *Proc. of Custom Int. Circ. Conf*, San Diego, California, pp. 24.3/1-4, May 1991.

[90] P. J. Crawley, and G. W. Roberts, "Switched-Current Sigma-Delta Modulation for A/D Conversion", *Proc. of Int. Symp. Circuits And Systems (ISCAS)*, San Diego, California, pp. 1320-1323, May 1992, IEEE.

[91] M. Bracey, W. Redman-White, and J. B. Hughes, "A Switched-Current Sigma Delta Converter for Direct Photodiode Interfacing", *Proc. of Int. Symp. Circuits And Systems (ISCAS)*, London, UK, pp. 4.287-4.290, May 1994, IEEE.

[92] S. Lindfors, K. Halonen, and J. Riihiaho "A Current Mode $\Sigma\Delta$-Modulator Based on the S^2I Error Compensation Technique", *Proceedings of ECCTD 95*, pp. 517-520, 1995.

[93] J. Nedved, J. Vanneuville, D. Gevaert, and J. Sevenhans, "A Transistor-Only Switched Current Sigma-Delta A/D Converter for a CMOS Speech CODEC", *IEEE J. Solid-State Circ.*, Vol. 30, No. 7, pp. 819-822, July 1995.

[94] N. Tan, and S. Eriksson, "A Low-Voltage Switched-Current Delta-Sigma Modulator", *IEEE J. Solid-State Circ.*, Vol. 30, No. 5, pp. 599-603, May 1995.

[95] N. Tan, "Fourth-Order SI Delta-Sigma Modulators for High-Frequency Applications", *Electron. Lett.*, Vol. 31, No. 5, pp. 333-334, Mar. 1995.

[96] S. V. Pattamatta, P. Manapragada, V. Dalal, and R. Schreier, "A Switched-Current Bandpass Delta-Sigma Modulator", *Proc. of Int. Symp. Circuits And Systems (ISCAS)*, London, UK, pp. 5.477-5.480, May 1994, IEEE.

[97] J. M. de la Rosa, B. Perez-Verdu, F. Medeiro, and A. Rodriguez-Vazquez, "CMOS Fully-Differential Bandpass Sigma Delta Modulator using Switched-Current Circuits", *Electron. Lett.*, Vol. 32, No. 3, pp. 156-157, Feb. 1996.

[98] J. B. Silva, C. A. Leme, and J. E. Franca, "A Fully-Differential, Self-Calibrated Switched-Current Delta Sigma Modulator", *Proceedings of IEEE Midwest Symp. Circ. Syst.*, Ames, Iowa, Vol. 2, pp. 1050-1053, Aug. 1996, IEEE.

[99] P. Simek, and V. Musil, "BIST for SI Sigma-Delta Analogue Front End", *Proc. of Int. Symp. Circuits And Systems (ISCAS)*, Hong Kong, Vol. 4, pp. 2729-2732, June 1997, IEEE.

[100] N. Tan, G. Amozandeh, A. Olson, and H. Stenström, "Current-Scaling Technique for High Dynamic Range Switched-Current Delta-Sigma Modulators", *Electron. Lett.*, Vol. 32, No. 15, pp. 1331-1332, July 1996.

[101] L. Quiquerez, and A. Kaiser, "Advanced Architectures for Current Memory Sigma-Delta Modulators", *Proc. of Int. Symp. Circuits And Systems (ISCAS)*, Hong Kong, Vol. 1, pp. 473-476, June 1997, IEEE.

[102] I. H. H. Jørgensen, and G. Bogason, "Optimization and Design of a Low Power Switched Current A/D- Sigma Delta -Modulator for Voice Band Applications", *Analog Integrated Circuits and Signal Processing*, Vol. 17, No. 3, pp. 221-247, Nov. 1998.

PART II

SI CIRCUIT DESIGN ISSUES

Chapter 4

Clock-Feedthrough Compensated First-Generation SI Circuit Design[1]

In this chapter we discuss the clock-feedthrough problem in switched-current circuits. We will focus on a clock-feedthrough compensated first-generation SI memory cell that ideally cancels both constant and signal-dependent clock-feedthrough. It is shown how to optimize the memory cell performance according to a general cost function. A test circuit was implemented in a single-poly CMOS process. Measured total harmonic distortion of the memory cell is less than -65 dB when optimized for low-power.

4.1 Introduction

A serious problem in SI circuits is switch-induced errors or *clock-feedthrough* (CFT). Without efficient CFT cancellation it is not possible to realize high-speed SI circuits with acceptable accuracy, and a number of solutions have been proposed, as shown in chapter 2. SI circuits also suffer from other non-ideal effects, such as *channel-length modulation* and *mismatch*. Theoretically, second-generation memory cells have no mismatch problem, and are in that sense advantageous over first-generation memory cells. This advantage is destroyed in general SI circuits, however, because current mirrors are always needed to generate multiple outputs. As will be shown in section 4.2, CFT also gives rise to an effective gain error,

---

[1] B. Jonsson, and N. Tan, "Clock-Feedthrough Compensated First-Generation SI Circuits and Systems", *Analog Integrated Circ. Signal Proc.*, Vol. 12, No. 3, pp. 201-210, Apr. 1997.

equivalent to that caused by transistor mismatch. Another drawback is the adverse effect of the transient current glitches in second-generation SI circuits. Second-generation SI circuits therefore have higher THD than first-generation circuits, as shown in [1]. In many applications, distortion is a more serious problem than linear gain errors. Therefore first-generation SI circuits with improved CFT cancellation was used throughout this work.

Section 4.2 describes the CFT problem and summarizes its modeling. An efficient CFT reduction scheme for first-generation SI memory cells giving low THD is also presented. How to optimize the CFT compensated memory cell according to an arbitrary goal function is shown in section 4.3, while chip measurement results are presented in section 4.4.

4.2 Clock-feedthrough compensation in SI circuits

Since clock-feedthrough is inherent in MOS switches, a thorough understanding and a careful modeling are essential in order to efficiently compensate for the CFT errors introduced. Because of the complexity introduced by detailed models, there is also a need for easy-to-use models that capture the main sources of CFT error. Publications dealing with the CFT problem in SI circuits [2-5], seem to favor model simplicity, rather than in-detail accuracy. In [6], however, the analysis is more in line with general results published on charge injection in MOS switches, e.g., [7-8]. A brief explanation of the CFT problem and the most frequently used CFT model are given below.

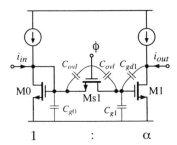

Fig. 4.1 Switched-current memory cell with parasitic capacitances.

4.2.1 Clock-feedthrough modeling

Consider the first-generation SI memory cell shown in Fig. 4.1. As the switch transistor Ms1 is turned off, charge is injected to the gate of M1 from the parasitic capacitances shown and from the conducting channel of Ms1.

The result is that an error voltage, V_{CFT}, is added to v_{GS1}, thus causing an error in the sampled current. The amount of charge injected can only be calculated using numerical methods [7-9]. It depends strongly upon the relative speed of signal transients, as well as device geometry. Closed form expressions can be derived if the switching is "slow" or "fast". The definition of slow and fast switching is found in [7-8]. Using the results valid for fast switching, the error voltage can be written as

$$V_{CFT} \cong \frac{3}{2} \cdot \frac{\Delta\phi W_{s1}\left(\frac{\eta}{2}L_{s1} + LD\right)}{W_1 L_1} \tag{4.1}$$

where

$$\Delta\phi = \phi_{low} - \phi_{high}$$
$$\eta = \frac{\phi_{high} - v_{GS0}}{|\Delta\phi|} \tag{4.2}$$

and LD is the lateral diffusion length.

4.2.2 Large-signal CFT error current

The sampling error caused by CFT can be calculated using large-signal analysis. Consider again the first-generation SI memory cell in Fig. 4.1. Assuming that M0 and M1 are operated in the saturation region, that the channel length modulation is negligible, and that M0 and M1 have identical processing parameters, the drain currents can be written as in Eqs. 4.3 and 4.4.

$$i_{D0} = I + i_{in} = \frac{\beta_0}{2}(v_{GS0} - V_T)^2 \tag{4.3}$$

$$i_{D1} = \alpha I + i_{out} = \alpha\frac{\beta_0}{2}(v_{GS0} - V_T + V_{CFT})^2 \tag{4.4}$$

Solving for i_{out} yields

$$i_{out} = i_{D1} - \alpha I = \alpha\left[i_{in} + V_{CFT}\sqrt{2\beta_0(I + i_{in})} + \frac{\beta_0}{2}V_{CFT}^2\right] \tag{4.5}$$

It is seen in Eq. 4.5 that V_{CFT} causes a nonlinear, signal-dependent error, and a constant offset. The square-root term normally dominates the error and gives rise to a constant offset, a change in the small-signal gain, and harmonic distortion. If we compare the current transfer function $i_{out}(i_{in})$ according to Eq. 4.5 with the ideal $i_{out} = \alpha\, i_{in}$, we conclude that the offset error caused by CFT is

$$e_{off} = \alpha \left[V_{CFT}\sqrt{2\beta_0 I} + \frac{\beta_0}{2} V_{CFT}^2 \right] \approx \alpha\, V_{CFT}\sqrt{2\beta_0 I} \qquad (4.6)$$

and the effective small-signal current gain, including CFT, is

$$\alpha_{eff} = \left(\frac{\partial i_{out}}{\partial i_{in}} \right)_{i_{in}=0} = \alpha \left[1 + V_{CFT}\sqrt{\frac{\beta_0}{2I}} \right] = \alpha \left[1 + e_\alpha \right] \qquad (4.7)$$

The gain error, e_α, is approximately -1 % with the transistor dimensions used in this chapter. It should be noted that Eq. 4.7 is valid for second-generation SI circuits as well, by choosing $\alpha = 1$. Thus, uncompensated second-generation circuits possess a gain error, similar to that caused by device mismatch in first-generation SI circuits, and therefore $\alpha_{eff} \neq 1$, contrary to what is often stated in the literature. They also have a fairly large THD. The performance of second-generation circuits is improved if S^2I or S^nI techniques are used [10-11]. In this chapter, however, we demonstrate the use of *first*-generation SI circuits with very small CFT error and THD.

4.2.3 Complete CFT cancellation

Assume that a memory cell output branch with gain α is realized by letting M1 have $W_1 = \alpha W_0$ and $L_1 = L_0$. Let V_C denote the CFT voltage when $\alpha = 1$. According to Eq. 4.1, the CFT voltage at the gate of transistor M1 is then

$$V_{CFT}(\alpha) = V_C/\alpha \qquad (4.8)$$

and Eq. 4.5 can be rewritten as

$$i_{out} = \alpha\, i_{in} + V_C\sqrt{2\beta_0(I+i_{in})} + \left(\frac{1}{\alpha}\right)\frac{\beta_0}{2} V_C^2 \qquad (4.9)$$

Note that the input signal dependent error term does **not** depend on α_i. This implies that it is always canceled when the output is formed as a difference between two output branches. This was proposed in [3], with $\alpha_1 = 1$ and $\alpha_2 = 2$, leaving only a constant CFT term uncompensated for. That circuit will be referred to as *Fiez memory cell* (FMC) in this chapter. A new CFT cancellation scheme for first-generation SI circuits was proposed in [12]. Ideally, it cancels *both* constant and signal-dependent CFT, thereby achieving *complete CFT cancellation*. Instead of using *two* output branches, as in the FMC, the *new memory cell*, (NMC), uses *four*, so that the total output is

$$i_{out} = (i_{o4} - i_{o3}) + (i_{o2} - i_{o1}) \tag{4.10}$$

The total output signal from the NMC is then

$$i_{out} = \left[(\alpha_4 - \alpha_3) + (\alpha_2 - \alpha_1)\right] i_{in} + \\ + \left[\left(\frac{1}{\alpha_4} - \frac{1}{\alpha_3}\right) + \left(\frac{1}{\alpha_2} - \frac{1}{\alpha_1}\right)\right] \frac{\beta_0}{2} V_C^2 \tag{4.11}$$

Unity gain memory cells will require $(\alpha_4 - \alpha_3) + (\alpha_2 - \alpha_1) = 1$. The condition for complete cancellation of the remaining CFT term is

$$\left(\frac{1}{\alpha_4} - \frac{1}{\alpha_3}\right) + \left(\frac{1}{\alpha_2} - \frac{1}{\alpha_1}\right) = 0 \tag{4.12}$$

There is an infinite number of solutions to Eq. 4.12 with unity gain. Choosing $\alpha_1 = 0.5$ and $\alpha_2 = 0.4$ yields $\alpha_3 \approx 1.03$ and $\alpha_4 \approx 2.13$. The SI memory cell shown in Fig. 4.2 will theoretically cancel both constant and signal-dependent CFT, thus realizing a CFT-free current memory cell. Since Eq. 4.5 is only an approximation, the CFT will not be completely canceled. The simulation results, however, indicate a substantial reduction of the remaining CFT error, even when compared with the FMC.

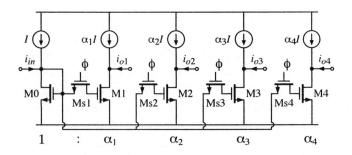

Fig. 4.2 Proposed CFT cancellation circuit. (NMC).
$\alpha_1 = 0.5$, $\alpha_2 = 0.4$, $\alpha_3 = 1.03$ and $\alpha_4 = 2.13$.

Fig. 4.3 shows the simulated CFT error current in the NMC compared to that of an uncompensated memory cell. The error is reduced with a factor 1000 or more, and is seen to become signal independent. This will greatly improve the THD. In fact, simulations reported in [3] showed that THD in the FMC was about 30 dB lower than in an uncompensated memory cell. Since Eq. 4.11 is valid for the FMC by choosing $\alpha_4 = \alpha_3 = 0$, $\alpha_2 = 2$ and $\alpha_1 = 1$, it is seen that the cancellation of the signal-dependent CFT errors in the NMC and FMC is essentially equal. Therefore the NMC is also expected to have similar THD performance.

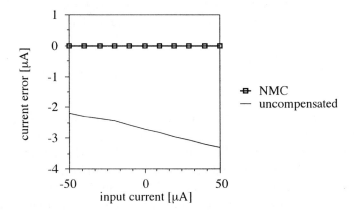

Fig. 4.3 Simulation data showing the CFT reduction achieved in the NMC as compared with an uncompensated memory cell.

4.2.4 Coefficient matching sensitivity

At a first glance, this technique seems to rely on very accurate coefficient matching, although that is not the case. Note that the square-root term is canceled by forming a difference – not by matching coefficients. Reduction of THD and most of the CFT does **not** depend on the successful realization of a certain α-value. The gain does, of course, and so does the cancellation of the constant CFT residue. It was seen in the simulations, that the coefficients could be truncated to one decimal, still maintaining superior CFT reduction. If unit-transistor techniques are used, sufficient matching is readily achieved [13-14]. It is sufficient to realize M0 with *ten* unit-transistors, each with $W_u = W_0/10$ and $L_u = L_0$. Each transistor Mi is then realized with $10 \cdot \alpha_i$ unit-transistors. Even with such a *coarse truncation*, the CFT is reduced by a factor *seven* or more, compared to the FMC, as seen in Fig. 4.4. Without truncation, the error is reduced by more than *ten* times throughout the input range.

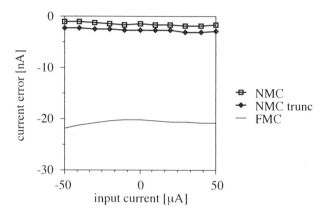

Fig. 4.4 Simulation data showing the CFT reduction achieved in the NMC with and without coefficient truncation. The FMC is included for comparison.

4.3 Memory cell design options

Complete cancellation is achieved when Eq. 4.12 and the gain requirement in Eq. 4.13 are satisfied.

$$\alpha = (\alpha_4 - \alpha_3) + (\alpha_2 - \alpha_1) \tag{4.13}$$

Since there are four variables and two equations, the designer will have some freedom that should be used to optimize the design. First it can be noted that current sample-and-hold circuits with arbitrary positive or negative gain can be realized with this circuit. It is obvious that the unity-gain solution can be scaled by α, but for a small or large gain it may be better to choose a new set of coefficients. The coefficients should not be chosen *too small*, because that would lead to large CFT voltages on the corresponding memory transistor. That can cause the transistor to turn off, or at least increase the harmonic distortion. On the other hand, if they are chosen *too large*, power and area are wasted, and the speed is reduced.

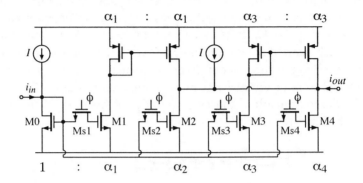

Fig. 4.5 Practical realization of the NMC.

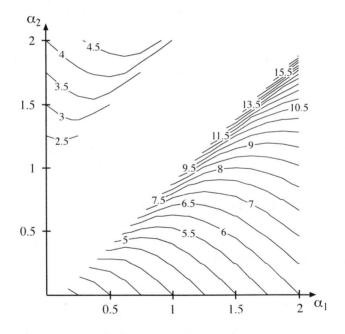

Fig. 4.6 Bias current in units of I, as a function of parameter choice.

4.3.1 Optimizing an arbitrary goal function

Any performance that depends on the α-coefficients can be optimized if a goal function, $G(\alpha_1, \alpha_2, \alpha_3, \alpha_4)$ can be specified. Two coefficients, e.g., α_1

and α_2, are chosen by the designer so that G is optimized. As an example, it is seen from Fig. 4.5 that the DC power dissipation is proportional to

$$I_{tot} = (1 + \alpha_1 + \alpha_2 + \alpha_3 + \alpha_4) \cdot I \qquad (4.14)$$

Thus, *minimum power dissipation* is achieved when the sum of coefficients is minimized. In Fig. 4.6, the total bias current is plotted vs. α_1 and α_2. A near-optimal solution, with reasonable coefficient values, is to choose $\alpha_1 = 0.3$ and $\alpha_2 = 1.4$ giving $\alpha_3 \approx 0.25$ and $\alpha_4 \approx 0.15$ [12]. A successful implementation using these values, is presented in this chapter. THD < -65 dB was measured. The power dissipation becomes $3.1 \cdot I$, which should be compared to $4 \cdot I$ for the standard implementation of the FMC. The FMC could however be optimized in a similar way, by shifting the coefficient values to $\alpha_1 = 0.5$ and $\alpha_2 = 1.5$ for example.

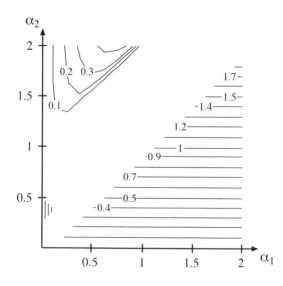

Fig. 4.7 Smallest coefficient value as a function of parameter choice.

4.3.2 Minimizing CFT

Ideally, *all* CFT is canceled in the NMC if the simple MOS equation is valid. Since that is not the case, there will still be some CFT. The largest error occurs in the smallest memory transistor. Thus, the CFT residue mainly depends on the size of the smallest transistor. In order to minimize CFT, one should maximize $Min\{\alpha_i\}$ while keeping $Max\{\alpha_i\}$ low enough to meet speed requirements. The goal function $Min\{\alpha_i\}$ is plotted in Fig. 4.7. The simulations in the previous section were done with $\alpha_1 = 0.5$ and $\alpha_2 = 0.4$ yields $\alpha_3 = 1.0$ and $\alpha_4 = 2.1$, which is a reasonable trade-off between accuracy, speed and power dissipation.

4.4 Experimental results[2]

The memory cell shown in Fig. 4.5 was implemented in a 0.8 μm digital CMOS process. Memory cells were optimized for low-power as shown previously. The bias current I was set to 100 μA. High-swing cascodes were employed to increase the input-output conductance ratio. The memory cell output power spectrum is shown in Fig. 4.8. With $f_c = 1$ MHz, $f_{in} = 10$ kHz, and $i_{in} = 10$ μA rms, the measured harmonic distortion of the memory cell is less than -65 dB (0.056 %).

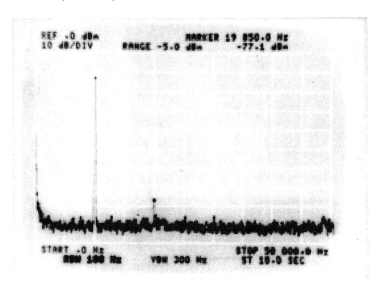

Fig. 4.8 Memory cell output power spectrum. Sampling rate 1 MHz, input signal 10 μA rms, 10 kHz. THD < -65 dB.

4.5 Summary

In this chapter we have discussed the clock-feedthrough problem in SI circuits and presented a clock-feedthrough cancellation scheme that ideally removes all clock-feedthrough in first-generation SI circuits. Its performance was carefully derived and analyzed, showing coefficient matching sensitivity

---

[2] The measurements were done by Dr. Nianxiong Tan.

to be moderate. A general method to optimize the memory cell performance for a variety of goal functions was illustrated. Measurement results showed that THD < - 65 dB, when optimized for power dissipation.

REFERENCES

[1] P. M. Sinn, and G. W. Roberts, "A comparison of first and second generation switched-current cells", in *Proc. of Int. Symp. Circ. Syst.*, London, UK, May 1994, Vol. 5, pp. 301-304.

[2] H. C. Yang, T. S. Fiez, and D. J. Allstot, "Current-Feedthrough Effects and Cancellation Techniques in Switched-Current Circuits", in *Proc. of Int. Symp. Circ. Syst.*, New Orleans, May 1990, pp. 3186-3188.

[3] T. S. Fiez, D. J. Allstot, G. Liang, and P. Lao, "Signal-Dependent Clock-Feedthrough Cancellation in Switched-Current circuits", in *Proc. of China 1991 Int. Conf. Circ. Syst.,*, Shenzhen, China, June 1991, pp. 785-788.

[4] R. T. Baird, T. S. Fiez, D. J. Allstot, "Speed and Accuracy Considerations in Switched-Current Circuits", in *Proc. of Int. Symp. Circ. Syst.*, Singapore, June 1991 pp. 1809-1812.

[5] M. Song, Y. Lee, and W. Kim, "A Clock Feedthrough Reduction Circuit for Switched-Current Systems", *IEEE J. Solid-State Circ.*, Vol. 28, No. 2, pp. 133-137, Feb. 1993.

[6] D. Macq, and P. Jespers, "Charge Injection in Current-Copier Cells", *Electron. Lett.*, Vol. 24, No. 21, pp. 1331-1332, Oct. 1988.

[7] G. Wegmann, E. A. Vittoz, and F. Rahali, "Charge Injection in Analog MOS Switches", *IEEE J. Solid-State Circ.*, Vol. SC-22, No. 6, pp. 1091-1097, Dec. 1987.

[8] J.-H. Shieh, M. Patil, and B. J. Sheu "Measurement and Analysis of Charge Injection in MOS Analog Switches", *IEEE J. Solid-State Circ.*, Vol. SC-22, No. 2, pp. 277-281, Apr. 1987.

[9] J. R. Burns, "Large-Signal Transit-Time Effects in the MOS-Transistor", *RCA Rev.*, vol. 15, pp. 14-35, Mar. 1969.

[10] J. B. Hughes, and K. W. Moulding, "S^2I: A Two-Step Approach to Switched-Currents", in *Proc. of Int. Symp. Circ. Syst.*, Chicago, Illinois, May 1993, pp. 1235-1238.

[11] C. Toumazou, and S. Xiao, "n-step charge injection cancellation scheme for very accurate switched current circuits", *Electron. Lett.*, Vol. 30, No. 9, pp. 680-681, Apr. 1994.

[12] B. Jonsson, and S. Eriksson, "New Clock-Feedthrough Compensation Scheme for Switched-Current Circuits", *Electron. Lett.*, Vol. 29, No. 16, pp. 1446-1447, Aug. 1993.

[13] B. Jonsson "Applications of the Switched-Current Technique", *Thesis* No. 458, LiU-Tek-Lic-1994:44, Linköping University, Sweden, 1994

[14] P. E. Allen, and D. R. Holberg, *CMOS Analog Circuit Design*, Holt, Rinehart, and Winston, Inc.: New York, 1987.

Chapter 5

Sampling Time Uncertainty[1]

Random and signal dependent sampling time uncertainty in high-speed switched-current circuits are analyzed in this chapter, and a comparison with voltage-mode sampling is made. The similarity of the two techniques is shown as well as the fact that the usually lower voltage swing in switched-current circuits, makes them less sensitive to the signal dependent switch-off time of the sampling switch. Derivations and simulation results showing the effects of clock phase-noise, additive clock driver noise, and signal-dependent sampling time uncertainty are included. Reduction of signal-dependent jitter errors by using fully-differential switched-current sampling is also illustrated.

5.1   Introduction

An input sampling circuit imposes a fundamental limit to the quality of a signal-processing or data-acquisition system. Any error introduced at this stage is directly added to the input signal and can not be compensated for in later processing stages. In particular, an accurately defined sampling time is essential for a high-speed, high-resolution system. It can be shown that to sample a 20 MHz sinusoidal with +/- 0.5 LSB error at 12-b resolution, timing errors are required to be less than +/- 2 ps [1]. As signal frequencies and resolution demands are increased, timing is becoming a major design issue in sampling circuits. Random sampling time uncertainty, or *jitter*, due to various noise-sources occurs as noise in the output. Signal-dependent

---

[1] B. E. Jonsson, "Sampling Jitter in High-Speed SI Circuits", *Proc. of Int. Symp. Circuits And Systems (ISCAS)*, Monterey, California, Vol. 1, pp. 524-526, May 1998, IEEE.

sampling jitter, caused by the variation of the threshold voltage in sampling switches increases the harmonic distortion. It will be shown in this chapter that switched-current (SI) circuits benefit from having lower voltage swing and that, by using fully-differential techniques, distortion can be further reduced, thus becoming negligible.

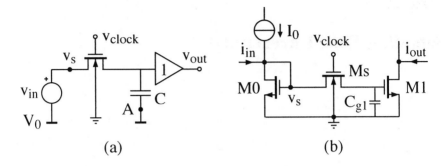

Fig. 5.1 Simplified sampling circuits: (a) voltage-mode. (b) current-mode.

5.2 Sampling time uncertainty

Simplified voltage- and current-mode sampling circuits are shown in Fig. 5.1 a and b respectively. The voltage-mode circuit samples the input voltage v_{in} onto the sampling capacitor C through the sampling switch Ms. The current-mode circuit samples the drain current in M0, $i_{D0} = i_{in} + I_0$, by sampling the gate source voltage v_{GS0} onto the gate capacitance of M1, C_{g1} [2]. In both circuits sampling occurs at $t = t_s$ when $v_{clock} = v_{off}$. The switch-off voltage v_{off} is

$$v_{off} = v_s + V_{T0} + \gamma \left(\sqrt{\phi + v_s} - \sqrt{\phi} \right) \tag{5.1}$$

where V_{T0}, γ and ϕ are the MOS technology parameters that determines the threshold voltage V_T of the sampling switch. The source voltage v_s is

$$v_s = v_{in} + V_0 \tag{5.2}$$

$$v_s = V_{T0} + \sqrt{\frac{2(I_0 + i_{in})}{K' W_0 / L_0}} \tag{5.3}$$

Sampling time uncertainty

for the voltage- and current-mode circuit respectively. Consider then the clock generator model shown in Fig. 5.2 a. A reference oscillator signal is fed to a comparator that detects zero-crossings. Any noise in the oscillator will cause the comparator to detect zero-crossings with a random time jitter, t_j. This causes a shift of the clock-edge by the same amount, as shown in Fig. 5.2 b. Noise generated or picked up by the clock-shaping circuit or subsequent clock drivers will add to the timing uncertainty. It is represented by the voltage v_n, added to v_{clock}. If a clock transition starts at $t = t_k + t_j$, then sampling occurs at

$$t_s \approx t_k + t_j + \frac{V_{DD} + v_n - v_{off}(t_k + t_j)}{a + \left(\frac{\partial v_{off}}{\partial t}\right)_{t=t_j+t_r}} \quad (5.4)$$

where normally

$$a = \frac{V_{DD}}{t_f} \gg \left|\frac{\partial v_{off}}{\partial t}\right| \quad (5.5)$$

and t_f is the clock fall time. Thus, the deviation from the nominal sampling time t_k can be expressed as

$$\Delta t_s \approx t_j + t_f \left(1 + \frac{v_n}{V_{DD}} - \frac{v_{off}}{V_{DD}}\right) = t_f + t_n + t_{in} \quad (5.6)$$

where

$$t_n = t_j + t_f \frac{v_n}{V_{DD}} \quad (5.7)$$

$$t_{in} = -t_f \frac{v_{off}}{V_{DD}} \quad (5.8)$$

are the random and input-dependent errors respectively. The random error t_n represents a random displacement of t_s that is only dependent on the clock signal and not on the sampling circuit or the signal level. Therefore SNR is degraded equally in both type of circuits. The input signal dependent error t_{in} causes harmonic distortion rather than noise. Its magnitude is a function of the source voltage v_s of the sampling switch Ms1. Therefore the sampling circuit that has the lowest voltage swing at this node will have the lowest THD. The dynamic range in voltage-mode circuits is directly proportional to the voltage swing, and therefore it is usually set as high as possible without compromising other design goals. In current-mode circuits the current can be increased without changing the voltage swing. Normally, the voltage swing is lower in a current-mode circuit [3], and therefore the effect of t_{in} is less obvious.

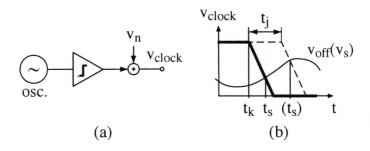

Fig. 5.2 Sampling clock: (a) Clock generator model. (b) Sampling time instance with and without random jitter.

5.3 Reducing sampling time uncertainty

5.3.1 Random jitter

Equation 5.6 shows that the effect of all error sources except clock source jitter can be reduced by decreasing the clock fall time. According to Eq. 5.7 the jitter of the reference oscillator dominates over the noise added to the clock if

$$t_f < \frac{t_j}{(v_n/V_{DD})} \tag{5.9}$$

If $t_j = 1$ ps, $v_n = 50$ mV, and $V_{DD} = 5$ V, then t_f should be less than 0.1 ns. This can also be seen from the MATLAB™ simulations shown in Fig. 5.3. A 21 MHz sinusoidal input was sampled at 50 MHz, with t_j, v_n, and V_{DD} as above. Equations 5.1 – 5.4 were used to model the sampling process. It is seen that, for $t_f < 0.1$ ns, SNR approaches a fixed level set by the oscillator jitter. With $t_f > 0.1$ ns, the noise level becomes proportional to t_f.

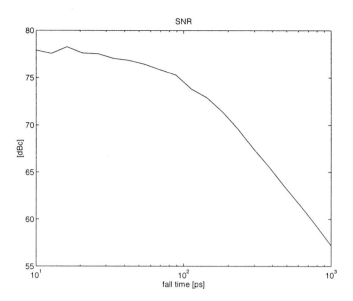

Fig. 5.3 MATLAB simulation of SNR vs. clock fall time in voltage- and current-mode sample-and-holds. Clock source jitter $t_j = 1$ ps rms, and $v_n = 50$ mV rms. When $t_f < 0.1$ ns, SNR is dominated by the clock source.

5.3.2 Signal dependent jitter

A well known technique to avoid signal dependent errors is available in *voltage-mode* sampling. A second sampling switch Ms2 can be inserted at "A" in the circuit in Fig 1 a. If Ms2 turns off slightly before Ms, the sampling time instance is determined by Ms2. Since the source of Ms2 is at AC ground, there will be no signal dependent errors. Most current-mode circuits, however, cannot utilize this technique[2] since the sampling capacitor is the gate-source parasitic, C_{g1}. It is seen from Eq. 5.8 that t_{in} can be reduced by decreasing either t_f or the variation in v_{off}. In SI circuits, the latter is done by increasing g_{m0}, thereby reducing the voltage swing v_s at the switch transistor. If this is not sufficient, fully-differential techniques can be used. In [4] it was shown that fully-differential sampling substantially reduces the effects of signal dependent sampling time uncertainty. When sampling a fully-differential signal the errors in the positive and negative signal cancel to the first order, giving up to 40 dB reduction of the signal dependent error, as shown for voltage-mode sampling. Similar results are achieved when *fully-differential current-mode sampling* is employed.

---

[2] Actually, a few *zero-voltage switching* SI circuits **have** been proposed [5-9], and this seems to be a growing field of research.

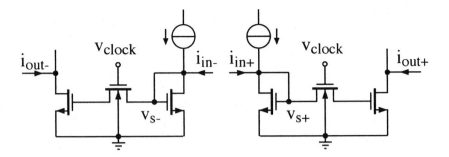

Fig. 5.4 Fully-differential current-mode sampling circuit.

A fully-differential current-mode sampling circuit is shown in Fig. 5.4. Let $i_{in+} = i_{in-} = 0$. Then $t_{s+} = t_{s-} = t_{s0}$ and both inputs are sampled at the same time. If $i_{in+} > 0$ then $t_{s+} < t_{s0}$, while $t_{s-} > t_{s0}$ by approximately the same amount τ so that

$$t_{s+} = t_{s0} - \tau$$
$$t_{s-} \approx t_{s0} + \tau \qquad (5.10)$$

The output signals thus become

$$i_{out+} \approx i_{in+} - \tau \cdot \frac{\partial i_{in+}}{\partial t}$$
$$i_{out-} \approx i_{in-} + \tau \cdot \frac{\partial i_{in-}}{\partial t} \qquad (5.11)$$

so that when the differential output i_{outd} is formed as

$$i_{outd} = i_{out+} - i_{out-} \approx i_{in+} - i_{in-} - \tau \cdot \frac{\partial i_{in+}}{\partial t} - \tau \cdot \frac{\partial i_{in-}}{\partial t} = i_{in+} - i_{in-} = i_{ind} \qquad (5.12)$$

it is found that the errors cancel to the first order. Because v_s is a more non-linear function of the input signal in SI circuits, the THD reduction is less than what was reported for voltage-mode circuits in [4]. Still, the simulations reported below indicate up to 20 dB reduction of THD. Let a sinusoidal input with $f_{in} = 21$ MHz be sampled at 50 MHz. Figure 5.5 shows THD vs. clock fall time simulated in MATLAB™ according to Eqs. 5.1 – 5.4. The voltage-mode input is +/- 1 V with $V_0 = 2.5$ V. The current-mode input is +/- 50 µA with $I_{bias} = 100$ µA into a current mirror with $W_0/L_0 = 32/3$ giving a voltage swing at v_s of about +/- 0.1 V. The current-mode sampling circuit has about 20 dB lower THD than the simple voltage-mode circuit included for comparison. This is *only* because of the lower swing. If the v_s swing is set equal in both circuits, the THD is almost identical. A more important result is that, by replacing the single-ended circuit with the fully-differential circuit in Fig. 5.4, THD can be reduced by **20 dB**.

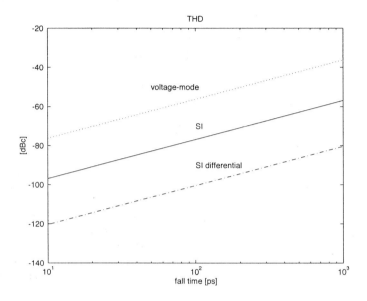

Fig. 5.5 MATLAB simulation of THD vs. clock fall time in voltage-mode (dotted), SI (solid) and fully-differential SI (dash-dotted) sample-and-holds. Signal swing is 2 V and 100 µA respectively, and f_{in} = 21 MHz.

5.4 Summary

Sampling time uncertainty in switched-current circuits was analyzed. It was shown that voltage- and current-mode sampling circuits have identical sensitivity to random clock jitter, but that current-mode circuits have the advantage of lower voltage swing. Therefore the signal-dependent sampling errors due to varying switch-off thresholds are less. Also, fully-differential sampling was seen to significantly reduce the effects of signal-dependent sampling errors.

REFERENCES

[1]   R. v. D. Plassche, *Integrated Analog-to-Digital and Digital-to-Analog Converters*, pp. 6-9, Kluwer Academic Publishers, 1994.

[2]   J. B. Hughes, I. C. Macbeth, D. M. Patullo, "Switched Currents - a New Technique for Analog Sampled-Data Signal Processing", *Proceedings of ISCAS 89*, Portland, Oregon, pp. 1584-1587, May 1989, IEEE.

[3]   C. Toumazou, F. J. Lidgey, and D. G. Haigh (Ed.), *Analogue IC design: the current-mode approach*, pp. 421-422, IEE Circuits and Systems Series 2, 1990.

[4]   B. Jonsson, S. Signell, H. Stenström, and N. Tan, "Distortion in sampling", *Proceedings of ISCAS 97*, Hong Kong, pp. 445-448, Jun. 1997, IEEE.

[5]   D. G. Nairn, "Zero-Voltage Switching in Switched-Current Circuits", *Proc. of Int. Symp. Circuits And Systems (ISCAS)*, London, UK, pp. 5.289-5.292, May 1994, IEEE.

[6]   P. Shah, and C. Toumazou, "A New High Speed Low Distortion Switched-Current Cell", *Proceedings of ISCAS 96*, Atlanta, Georgia, Vol. 1, pp. 421-424, May 1996, IEEE.

[7]   H. Ishii, S. Takagi, and N. Fujii, "Switched Current Circuit using Fixed Gate Potential and Automatic Tuning Circuit", *Electrical Engineering in Japan*, Vol. 126, No. 3, pp. 21-29, Feb. 1999.

[8]   A. Worapishet, J. B. Hughes, and C. Toumazou, "Class AB Technique for High Performance Switched-Current Memory Cells", *Proc. of 1999 Int. Symp. Circuits and Systems*, Orlando, Florida, Vol. 2, pp. 456-459, May 1999, IEEE.

[9]   J. M. Martins, and V. F. Dias, "Very Low-Distortion Fully Differential Switched-Current Memory Cell", *IEEE Trans. on CAS-II*, Vol. 46, No. 5, pp. 640-643, May 1999.

Chapter 6

Design of Power Supply Wires[1]

In this chapter resistive voltage drop on power supply lines is analyzed with respect to its effect on analog circuit performance. It is shown that proper power supply interconnection design is essential for low-distortion, low-offset, current-mode circuits. The results are particularly applicable to the design of current-mode A/D and D/A converters. Formulas suitable for hand-calculations are derived and design examples are included.

6.1 Introduction

In high-precision current-mode circuits, such as A/D and D/A converters, a very accurate matching of MOS transistor drain currents is necessary. Resistive voltage drops along power supply lines may range from a few microvolts to hundreds of millivolts. This voltage drop will cause an offset in V_{GS}, resulting in drain current mismatch and harmonic distortion. The acceptable amount of mismatch and distortion depends on the application. Bias currents for current mirrors and SI memory cells may be allowed to vary as much as 5-10% with negligible performance loss as long as the local mismatch within each mirror is small. When accurate multiple copies of an analog input signal or reference are generated by current mirrors, the situation turns out somewhat different. It will be shown in this chapter that even in a small circuit, the currents flowing through the devices can cause a supply wire voltage drop large enough to degrade the performance. A few

---

[1] B. E. Jonsson, "Design of Power Supply Lines in High-Performance SI and Current-Mode Circuits", *Proc. of 15th NORCHIP Conf.*, Tallinn, Estonia, pp. 245-250, Nov. 1997, IEEE.

92 Chapter 6

design examples will illustrate the importance of interconnection design in current-mode circuits.

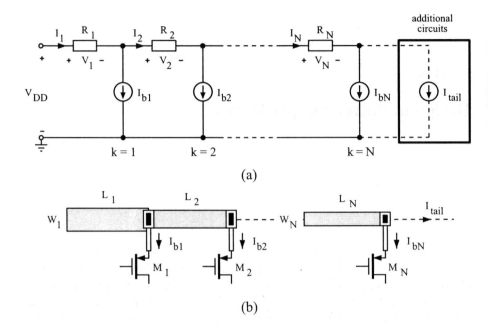

Fig. 6.1 (a) General circuit schematics showing the voltage drop caused by N subcircuits connected to the same V_{DD} wire. Additional circuits loading the same power supply line is represented by the sum of their currents, I_{tail}. (b) Supply wire geometry definitions and parts of a typical SI or current-mode circuit shown.

6.2 Voltage drop on power supply wires

Assume that the circuit of interest can be simplified to the circuit shown in Fig. 6.1 a. The current in each subcircuit branch is represented by a current source, I_{bi}. V_{DD} and ground is connected to one end of the circuit. This is typical for many VLSI circuits. The interconnection resistance and the voltage drop of each wire segment are denoted by R_i and V_i respectively. In some situations it is only the first N currents that has to be designed for. Although the total current is significant for the result, it is sufficient to represent the terminating circuits with the sum, I_{tail}, of their currents. Voltage drop on the ground wire can be derived in a similar way and is therefore omitted for simplicity. Figure 6.1 b shows the definition of supply wire dimensions and a part of a typical CMOS current-mode circuit. It is seen from Fig. 6.1 a that the accumulated voltage drop at point k along the supply-wire is:

Design of power supply wires

$$V_{drop}(k) = \sum_{i=1}^{k} V_i = \sum_{i=1}^{k} R_i I_i = \sum_{i=1}^{k} R_i \left(I_{tail} + \sum_{j=i}^{N} I_{bj} \right) \tag{6.1}$$

Including the sheet resistance, R_{sh}, and geometric information according to Fig. 6.1 b we get a general expression for the voltage drop along the wire:

$$R_i = R_{sh} \frac{L_i}{W_i} \Rightarrow V_{drop}(k) = R_{sh} \sum_{i=1}^{k} \frac{L_i}{W_i} \left(I_{tail} + \sum_{j=i}^{N} I_{bj} \right) \tag{6.2}$$

A special case that can be useful for quick estimations is that of N evenly spaced identical branches, supplied with a single wire of constant width W and length L_{tot}. Many other circuits can be approximated by this special case, and since Eqs. 6.4 and 6.5 are suitable for hand calculations, the required wire width can quickly be estimated.

$$I_{bi} = I_b, \quad W_i = W, \quad L_i = \frac{L_{tot}}{N} \quad \forall\, i \ni \{1 \ldots N\} \tag{6.3}$$

$$V_{drop}(k) = R_{sh} \frac{L_{tot}}{W} \frac{k}{N} \left(I_{tail} + I_b \left[(N+1) - \frac{(k+1)}{2} \right] \right) \tag{6.4}$$

$$V_{drop}(N) = R_{sh} \frac{L_{tot}}{W} \left(I_{tail} + I_b \frac{(N+1)}{2} \right) \tag{6.5}$$

Considering the circuit shown in Fig. 6.1 b, the actual branch current I'_{bk} is related to the ideal current I_{bk} and the supply voltage drop as:

$$I'_{bk} \approx I_{bk} - \sqrt{2\beta_0 I_{bk}} \left[V_{drop}(k) \right] + \frac{\beta_0}{2} \left[V_{drop}(k) \right]^2 \tag{6.6}$$

Equation 6.6 is derived from the large signal MOS equations and is valid even when $V_{drop}(k)$ is large or if I_{bk} includes a signal, as long as all devices remain in saturation. If $V_{drop}(k)$ is small, and no signal is present in I_{bk} then:

$$I'_{bk} \approx I_{bk} - g_{mk} V_{drop}(k) \tag{6.7}$$

6.3 Design considerations

The following simplified design examples are taken from the 10-b, 32 MS/s, parallel pipelined A/D converter implementation that is described in chapter 11 of this book. In order to achieve a high sampling rate, it employs eight 4 MS/s ADCs working in parallel. It is natural from a layout point of view to place the ADCs in eight rows with nine bitcells, as shown in Fig. 6.2 b, each row with its power supplies fed from the MSB side. The total bias current for a single row is ~ 12.6 mA, so it will create a large DC voltage drop. In fact, large enough to shift the quiescent point of the last bitcells considerably, unless the power supply wires are properly designed. This is illustrated in the following section. Since the ADCs operate in parallel, a number of precisely matched copies of the input current is needed. It will also be seen below that the small current-copying circuit in Fig. 6.2 a is capable of generating enough voltage drop to corrupt its own operation.

6.3.1 Quiescent point shift

Consider the design example in Fig. 6.2. The load on each VDD line is nine bitcells approximated with $9 \cdot 14 = 126$ evenly spaced 100 µA PMOS current sources. Thus $N = 126$, $I_b = 100$ µA and the total current is 12.6 mA. If the maximum current density in a wire is 1 mA/µm, then W has to be at least 12.6 µm to meet that requirement. However, using Eq. 6.5 with $W = 12.6$ µm, $I_{tail} = 0$, $L_{tot} = 3000$ µm, and a worst case $R_{sh} = 0.08$ Ω/square, gives $V_{drop}(N) \approx 121$ mV.

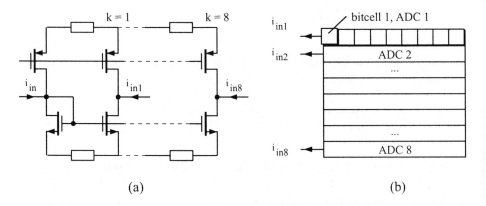

Fig. 6.2 Design example showing a 51.2 MS/s SI parallel pipelined A/D-converter. (a) Input current mirror generating eight copies of the input current, i_{in}. (b) ADC core block with eight sub converters operating in parallel.

Design of power supply wires 95

Circuit-level simulations showed $g_m \approx 400$ µA/V, and therefore the bias current in the last circuit will be reduced to $I'_{bN} \approx I_b - g_m V_{drop} = 51.6$ µA. Although $V_{drop}(N)$ in reality would be less, since most branch currents are now significantly less than their ideal values, we can see that there will be a *major shift in the operating point*. If the bias current is reduced by as much as 30 – 50 %, the maximum signal swing is reduced by the same amount, and therefore THD or noise will increase. Combining Eqs. 6.5 and 6.7 we get the wire width requirement given a certain deviation in bias current.

$$W \geq R_{sh} L_{tot} \frac{g_m}{I_b - I'_{bN}} \left(I_{tail} + I_b \frac{(N+1)}{2} \right) \tag{6.8}$$

With a tight tolerance for the bias current, the supply wire width will increase dramatically, as indicated in Table 6.1. A 10 µA reduction was considered acceptable in this particular design and thus $W = 61$ µm was used.

$I_{bN} - I'_{bN}$ [µA]	W_{VDD} [µm]
10	61
5	122
1	610

Table 6.1 Supply wire minimum width vs. bias tolerance.

6.3.2 Offset and distortion

Assume that eight copies of a +/- 50 µA full-scale (FS) input signal are generated by the NMOS current mirror in Fig. 6.2 a and distributed to each ADC row. Any mismatch between the copies will appear in the output signal as spurious frequencies. Thus we require the mismatch to be less than one LSB, which is -48 dBFS for 8-b and -60 dBFS for 10-b matching. Let $N = 8$, $I_{bias} = 100$ µA, $I_b = I_{bias} + i_{in}$, $g_{mN} = 570$ µA/V, and $L_{tot} = N \cdot 15$ µm. MATLAB™ simulation of the current mirrors according to Eq. 6.6 gives the THD and offset shown in Fig. 6.3. The simulation accounted for the fact that i_{in} and its replicas contributes significantly to V_{drop} on the NMOS side, and therefore harmonic distortion is further increased. It was also included that the effect of V_{DD} and ground voltage drops tend to cancel each other regarding the DC offset term in the output current. Simulations show that, to achieve 8 or 10-b matching, $W_{wire} \geq 7$ and 24 µm respectively. Offset dominates over harmonic distortion. Still we see that THD $\leq$ -70 dBc at full-scale input requires $W_{wire} \geq 8$ µm even for this small circuit.

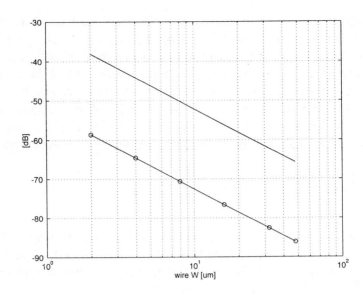

Fig. 6.3 MATLAB simulation of THD and offset in the 8th copy of i_{in}.
(o) THD [dBc], (—) DC offset [dBFS]

6.4 Summary

In this chapter we showed that supply line voltage drop has a significant impact on the performance of high-accuracy current-mode circuits such as A/D- and D/A converters. If the circuit would be ideal in every other sense, we can *still* not achieve even 8-b matching of drain currents if the voltage supply line is less than a certain *non-minimum* width. It was also shown that in a larger circuit, the quiescent currents may be shifted significantly. Therefore, careful design of V_{DD} and ground lines was found to be necessary for global as well as local drain current matching.

Chapter 7

SI Circuit Layout

7.1 Introduction

Most analog IC design includes careful design of the physical layout. The performance gained during the circuit design phase can easily be destroyed by a poor layout. As will be shown in this chapter, many switched-current designs have a simpler and much more regular structure than similar switched-capacitor circuits. By utilizing this regularity, and applying sound design rules, it is possible to shorten the layout design phase **and** maintain circuit performance. A further reduction of design time is possible if automatic layout generation is used. It will be shown in this chapter that SI circuits are well suited for layout automation. A simple CAD tool structure is suggested in the last section. The regular structure of SI circuits is helpful also when doing manual layout, and two layout styles suitable for manual and automatic layout are therefore described. Analog layout design has been thoroughly described elsewhere, i.e., [1-5], and a general treatment will not be included here. A few other authors have presented work regarding design automation for SI circuits: Hughes et. al. developed switched-current cells for design automation [6] and an automated design system for SI filters [7-8] using a *Cadence* framework, and O'Connor et. al. have developed a tool for automated design of switched-current cells [9].

7.2 Difference between SI and SC circuit layout

There are some important differences between SI and SC circuit layout. Consider the two simplified sampling circuits in Fig. 7.1. First it is noted that a switched-capacitor circuit includes a high-gain amplifier, which is *not* used in the switched-current circuit. Secondly, dedicated sampling capacitors (*C*) are used in the SC circuit, while the SI circuit samples on the parasitic gate capacitance (C_{gs}). These differences have implications on the layout design:

- The high-gain voltage amplifier is almost certain to have a more complex and irregular topology than the SI memory cell. Its actual size and aspect ratio is therefore more unpredictable than the size and aspect ratio of an SI memory cell.
- In the SC circuit, sampling capacitors need to be shaped and placed according to the size of the amplifier(s), while no capacitors are used in the SI circuit. This step can therefore be omitted.

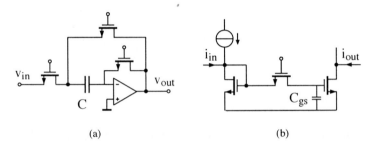

Fig. 7.1 Sampling circuits: (a) Switched-capacitor circuit. (b) Switched-current circuit. Sampling on parasitic gate capacitance.

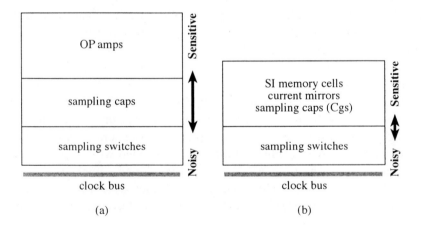

Fig. 7.2 Typical layout. (a) SC circuit. (b) SI circuit.

A typical SC circuit is laid out according to Fig. 7.2 a. The clock bus and the sampling switches are placed as far away as possible from the sensitive signal nodes at the OP-amps. The sampling capacitors are placed in between the switches and the amplifiers [3, 5]. A corresponding SI circuit layout is

SI circuit layout

shown in Fig. 7.2 b. The SI circuit layout becomes less complex already because of the absence of sampling capacitors. Taking into account that the remaining circuits in Fig. 7.2 b are essentially current mirrors and switch transistors, it is reasonable to assume that a majority of SI circuits have a simpler layout – more suitable for layout automation. This will be further developed in sections 7.4 and 7.5.

7.3 Floorplanning

Switched-current circuits could typically be used as interfacing circuits on a large mixed-signal chip. An important design issue is then to protect the sensitive analog parts from the influence of digital switching noise present in the substrate and on the power supplies. Power supply switching noise can be reduced by using separate analog and digital power supplies, while substrate noise is reduced by physically separating the sensitive analog parts from the noisy digital parts. The floorplanning example below is taken from the *dual parallel pipeline ADC* design described in chapter 11. Often, a "mixed-signal chip" is interpreted as something more than just an ADC on its own. In this case, the ADC itself has enough digital correction circuits and noisy output drivers to justify its use as a mixed-signal design example.

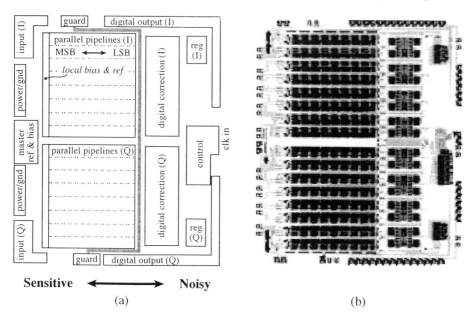

Fig. 7.3 Chip floorplanning for signal integrity. (a) Floorplan. (b) Chip layout.

The floorplan and the chip layout are shown in Fig 7.3 a and b. It is indicated how the amount of digital switching noise gradually increases to the right, and the sensitivity to noise increases towards the left side of the chip. The most sensitive parts are the input signal, master reference and bias circuits. These parts are therefore placed furthest to the left, as far away as possible from the digital output pads. Almost equally sensitive are the MSB stage and the local bias and reference generators in each pipeline. The pipelines are therefore placed in rows where the LSB stage is closest to the digital side of the chip. To improve substrate noise isolation a guard bar is laid out between the analog and digital parts, as indicated by the thick gray line in the floorplan. The noisy part of the chip is the area containing the digital output pads and the large drivers used for the 40-b internal clock bus. These circuits were given their own power supplies, and placed to the far right of the chip. Due to the large number of output pads it was unavoidable that some of them ended up close to the analog parts. The digital correction logic is less noisy than the pads and clock drivers, and therefore it was also given a power supply of its own. By doing so, the digital correction region of the chip will actually screen some of the substrate noise transmitted from the pads and clock drivers.

7.4 Layout styles

Layout design is usually simplified by the consistent use of design styles. The two layout styles described below were used for the circuits presented in this book. They are useful for manual **and** automatic layout design. In fact, by studying the proposed layout styles, the potential for automatic layout generation should become obvious. The styles are applicable with little or no modification to any basic 1st and 2nd generation SI circuit[1] using the low-swing cascode [10] shown in Fig. 7.4 a. More complex layout styles may be necessary for other classes of SI circuits.

7.4.1 Style I

One of the main advantages with the low-swing cascode is that all geometry scale linearly with the current gain[2] α in one direction, while remaining fixed in the other. Consider the simplified layout for cascode branches with $\alpha = 1$ and $\alpha = 1.5$, respectively, shown in Fig. 7.4 b. As seen from Fig 7.4 a, the width of all transistors were chosen as an integer multiple of the lower NMOS transistor. This is reflected in the layout as the 2, 3 and 6

---

[1] Including, for example, S^2I, and S^nI memory cells.
[2] For a definition of current gain, see chapter 4.

segments forming the other three transistors. Wiring to connect individual transistor segments has been excluded for simplicity. Although this choice of dimensions is not strictly necessary, it significantly improves area efficiency and matching, since devices can be packed more closely. The entire cascode branch is itself divided into a number of *unit segments*. Here, the number of segments was set to four, giving a gain resolution of 0.25. The actual number of segments is determined by the application. This layout style was selected for the filter implementation in chapter 8, using 25 segments to achieve a coefficient resolution of 0.04. As seen from Fig. 7.4 b, all transistor channels are oriented vertically, giving a fixed height. When α changes, only the number of unit segments is changed – proportional to α. Since each transistor segment is placed on a fixed pitch, cells can be *abutted*, which greatly simplifies the compilation of larger modules. Unit-segments from different cascode branches can be interdigitized in order to improve coefficient matching. Design rules normally require PMOS and NMOS transistors to be widely spaced. This requirement naturally leads to the *internal* horizontal *routing channel* apparent in Fig. 7.4 b. It is also possible to have a vertical routing channel between each cascode branch if necessary, or between each leaf-cell. Both channels can be used for *global* and *local* signal distribution, and their widths are easily changed according to the individual needs of each block or chip.

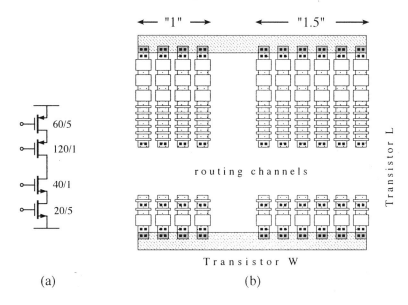

Fig. 7.4 Regular layout style I. (a) Unity-gain cascode branch. (b) Layout.

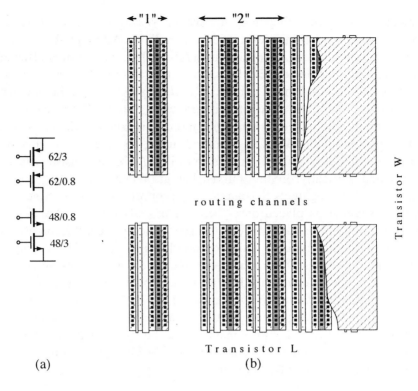

Fig. 7.5 Regular layout style II. (a) Unity-gain cascode branch. (b) Layout.

7.4.2 Style II

A slightly different layout style, shown in Fig. 7.5, was used for the A/D converter implementations in chapters 10 and 11. This second style is less useful for accurate realization of filter coefficients, but well suited for ADC circuits where the gain is usually 1, 2, 4, 8 ... etc. The transistor channels are oriented horizontally. This orientation gives the freedom to choose arbitrary widths for the N and PMOS transistors. For best area usage, the PMOS cascode has the same width as the PMOS current source, and the width of the NMOS cascode is the same as the NMOS memory transistor. Again, choosing these cascode sizes is not necessary but improves packing density. Another area-saving feature of style II is that power and ground wires can be layered on top of the transistors. Very wide supply wires can therefore be used without area penalty, as long as they match the transistor dimensions. Every cascoded transistor has its own substrate/well contact and via. The substrate and wells are therefore tightly coupled to the supplies, and the current flowing through the device is supplied through a short path having the lowest possible resistance. Due to the good substrate connection, and the shielding effect of the MET2 power supplies, this style is also believed to be less sensitive to external noise.

7.5 Layout automation

Manual layout is one of the major bottlenecks in the design process. Therefore a great deal of work has been focused on automatic layout generation, e.g., [11-12]. There are many obvious advantages with auto-generated IC layout. In commercial applications it will enable analog ICs with a short time-to-market. Furthermore, several major design changes, such as process parameter changes and resizing of building blocks, are quickly propagated to the final layout. Careful post-layout simulation is also more likely to be done on a large number of alternate designs, if the designer is not facing the tedious task of manual redesign. Since SI circuit performance depends largely on the memory cell, this is important in the search for an optimal design. It also opens up the possibility for mixed analog-digital designers who are inexperienced in analog design to include the analog interfaces in their digital design flow.

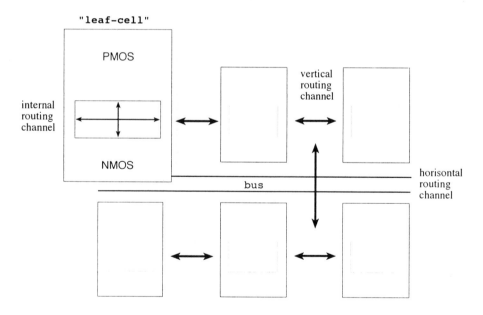

Fig. 7.6 Standard-cell layout style for SI and current-mode circuits

Leaf-cells designed according to the previously discussed styles can be placed in a *standard-cell manner* as shown in Fig. 7.6. Placement and routing can be done with commercial place-and-route tools. The simple hierarchical SI CAD tool structure shown in Fig. 7.7 was developed by the author in order to demonstrate the potential for automatic layout generation. Low level circuits such as current mirrors, 1-b DACs and current sample-and-holds are built from a limited set of fixed or parameterized primitives. Higher level modules like filters and Δ-Σ modulators are compiled from parameterized current mirrors, memory cells DACs and comparators. Design rules and parameters such as filter coefficients and geometry are determined in earlier stages of the design, and stored in files that are read by the module generators. By using technology dependent design rules as parameters, a high degree of portability is achieved. Module generators were written using the *L-language* [13] built in to the *GDT* layout design tool [14], but a similar tool could be developed in any design environment that supports scripting. The *Wave SI Filter* (WSIF) in chapter 8 was used to demonstrate the possibility to auto-generate SI systems [15-16]. The entire layout can be completely regenerated with a new set of design rules, filter coefficients and/or nominal transistor sizes within minutes. It was successfully proven during the actual design of the chip, when an unexpected change of design rules from 1.0 to 0.8 μm was announced only a few weeks before deadline.

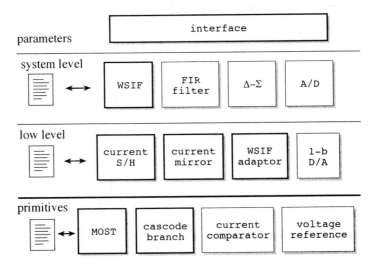

Fig. 7.7 Proposed SI CAD tool structure. Parts that were implemented have been highlighted.

REFERENCES

[1] M. Ismail, and T. Fiez, *Analog VLSI: Signal and Information processing*, McGraw-Hill, 1994.

[2] N. Tan, Switched-Current Design and Implementation of Oversampling A/D Converters, Kluwer, 1997.

[3] D. A. Johns, and K. Martin, *Analog Integrated Circuit Design*, Wiley, 1997.

[4] P. E. Allen, and D. R. Holberg, *CMOS Analog Circuit Design*, Holt, Rinehart, and Winston, Inc., New York, 1987.

[5] K. R. Laker, and W. M. C. Sansen, *Design of Analog Integrated Circuits and Systems*, McGraw-Hill, 1994.

[6] J. B Hughes, and K. W. Moulding, "Switched-Current Cells for Design Automation", *Proc. of Int. Symp. Circuits And Systems (ISCAS)*, Seattle, Washington, pp. 2067-2070, May 1995, IEEE.

[7] J. Richardson, J. B. Hughes, K. W. Moulding, M. Bracey, W. Redman-White, J. Bennett, and R. S. Soin, "An Integrated Design and Synthesis System for High Performance Switched-Current Analogue Filters", *Proceedings of ESSCIRC 95*, Lille, France, pp. 202-205, Sept. 1995.

[8] J. B. Hughes, K. W. Moulding, J. Richardson, J. Bennett, W. Redman-White, M. Bracey, and R. S. Soin, "Automated Design of Switched-Current Filters", *IEEE J. Solid-State Circ.*, Vol. 31, No. 7, pp. 898-907, July 1996.

[9] I. O'Connor, and A. Kaiser, "Automated Design of Switched-Current Cells", *Proc. of Custom Int. Circ. Conf*, Santa Clara, California, pp. 477-480, May 1998.

[10] J. B. Hughes, N. C. Bird, and I. C. Macbeth, "Switched Currents – A New Technique for Analog Sampled-Data Signal Processing", *Proc. of Int. Symp. Circuits And Systems (ISCAS)*, Portland, Oregon, pp. 1584-1587, May 1989, IEEE.

[11] J. D. Bruce, H. W. Li, M. J. Dallabetta, and R. J. Baker, "Analog Layout Using ALAS!", *IEEE J. Solid-State Circ.*, Vol. 31, No. 2, pp. 271-274, Feb. 1996.

[12] B. R. Owen, R. Duncan, S. Jantsi, C. Ouslis, S. Rezania, K. Martin, "BALLISTIC: An Analog Layout Language", *Proc. of Custom Int. Circ. Conf*, Santa Clara, California, pp. 41-41, May 1995.

[13] *L Language Reference*, Mentor Graphics Corporation, Wilsonville, OR.

[14] *GDT User's Manual*, Mentor Graphics Corporation, Wilsonville, OR.

[15] B. Jonsson, and S. Eriksson, "A Low Voltage Wave SI Filter Implementation Using Improved Delay Elements", *Proc. of Int. Symp. Circuits And Systems (ISCAS)*, London, UK, pp. 5.305-5.308, May 1994, IEEE.

[16] B. Jonsson, *Applications of the Switched-Current Technique*, Licentiate Thesis No. 458, Linköping University, Sweden, Oct. 1994.

PART III

SI CIRCUIT IMPLEMENTATION EXAMPLES

Chapter 8

A 3.3-V CMOS Wave SI Filter[1]

In this chapter a low voltage wave switched-current filter is presented. Circuit realizations of current-mode N-port adaptors and delay elements are given. The use of three-port adaptors allows realization of wave SI filters with transmission zeros. In order to enhance the performance, delay elements with an improved clock-feedthrough cancellation technique are used. Simulation results from a complete wave SI filter circuit operating with a 3.3 V power supply are included. A test filter has also been implemented in a standard CMOS process. The chosen filter and circuit structure is shown to be suitable for automatic layout generation.

8.1 Introduction

Filters are of fundamental importance in signal processing. Analog discrete-time filters can be realized using SI techniques, where the filter coefficients are determined by the relative size of MOS transistors [1]. In order to make accurate SI filters, one must choose a filter structure with low coefficient sensitivity. Wave digital filters are known to be stable filters with low sensitivity to coefficient variations [2]. In digital filters this relaxes the requirements on coefficient word length, thus reducing the chip area. In a corresponding analog sampled-data realization, low coefficient sensitivity leads to relaxed matching requirements. This is important since analog signal scaling cannot be done with arbitrary precision. In CMOS *switched current*

---

[1] B. Jonsson, and S. Eriksson, "A Low Voltage Wave SI Filter Implementation using Improved Delay Elements", *Proceedings of 1994 Int. Symp. Circuits and Systems*, London, UK, pp. 5.305-5.308, May 1994, IEEE.

(SI) filters, the scaling coefficients are determined by MOS transistor ratios, which are subject to random variations due to the fabrication process. In order to make accurate SI filters, it is important to choose a filter structure with low coefficient sensitivity. It is believed that the discrete-time wave filter approach is a good solution to this problem. Current-mode circuit realizations of two-port adaptors for wave SI filters were presented in [3]. Filters corresponding to a microwave prototype with cascaded unit elements were realized using a combination of such adaptors and current memory cells. A general current-mode circuit structure for N-port adaptors (where $N \geq 3$) was proposed by the author in [4] and is presented in this chapter. By using such adaptors, e.g. with $N = 3$, it is possible to realize filters with transmission zeros on the unit circle in the z-plane. One example of such filters is the elliptic filter. More generally, the number of possible filter realizations is greatly increased with the N-port adaptor available.

The simple and highly regular structure of the circuits presented in this chapter also indicates great possibilities for *automatic generation*. Analog integrated circuits generally require a very careful design to ensure proper operation. Building blocks such as the OP-amp are often built from a number of transistors with a large variation in geometry. Therefore they are not as suitable for automatic generation as digital circuits are. This leads to a time-consuming design, and the cost for the analog part of a mixed-mode chip may be large. Thus, analog circuits that are suitable for automatic layout generation are very interesting for commercial applications.

Clock feedthrough (CFT) errors are inherent in all CMOS SI circuits, and without some form of CFT compensation, SI circuits will never be able to meet high-performance signal processing specifications. In this chapter we therefore illustrate the benefits of an improved CFT-compensated SI memory cell, suitable for a wide range of signal processing applications.

8.2   Discrete-time wave filters

The now well-known *wave digital filter* (WDF) concept was originally introduced by Fettweis [5]. Numerous references dealing with WDF theory and applications can be found in the literature, e.g., [2]. The basic idea is to map passive *microwave reference filters*, which are known to be stable, onto discrete-time filters that simulate the wave propagation in the reference filters. A discrete-time wave filter has two principal building blocks: *delay elements* simulating the propagation delay in a wave-guide, and *adaptors*. Adaptors are used to simulate the reflections caused by a parallel- or series-connection of filter elements. A two-port adaptor realization simulating the effect of two parallel/series-connected components was proposed in [3]. The generic N-port adaptor was proposed by the author in [4]. It simulates the parallel/series-connection of an arbitrary number (N) of reference filter elements. Two-port and N-port parallel and series adaptor symbols are shown in Fig. 8.1. The corresponding adaptor equations are 8.1 a-c for the two-port adaptor, and 8.2 and 8.3 a-c for the N-port parallel and series adaptor respectively [2]. A_k and B_k are the waves respectively entering and leaving port k, and R_k is the *characteristic impedance* associated with port k.

$$B_1 = A_2 + \gamma(A_2 - A_1) \tag{8.1a}$$

$$B_2 = A_1 + \gamma(A_2 - A_1) \tag{8.1b}$$

$$\gamma = \frac{R_1 - R_2}{R_1 + R_2} \tag{8.1c}$$

$$A_0 = \sum_{k=1}^{N} \gamma_k A_k \tag{8.2a}$$

$$B_k = A_0 - A_k \tag{8.2b}$$

$$\gamma_k = \frac{2G_k}{\sum_{k=1}^{N} G_k} \quad \text{where} \quad G_k = \frac{1}{R_k} \tag{8.2c}$$

$$A_0 = \sum_{k=1}^{N} A_k \tag{8.3a}$$

$$B_k = A_k - \gamma_k A_0 \tag{8.3b}$$

$$\gamma_k = \frac{2R_k}{\sum_{k=1}^{N} R_k} \tag{8.3c}$$

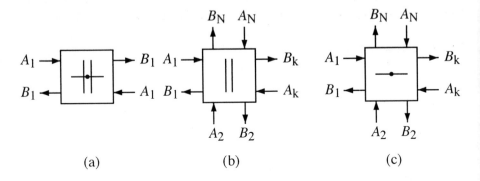

Fig. 8.1 Adaptor symbols. (a) Two-port. (b) N-port parallel. (b) N-port series.

8.3 Current-mode realization of N-port adaptors

In a *wave SI filter* (WSIF), currents represent the wave variables. Three operations are needed to realize an adaptor: *scaling, sign inversion* and *addition*. Sign inversion is inherent in the current mirror, and scaling is determined by the mirror ratio [6]. Addition of currents is based on Kirchhoff's current law, and implemented simply by wiring outputs together. Thus, the two-port and the N-port parallel and series adaptor described by equations 8.1 – 8.3 can be realized with current mirrors according to the circuit in Fig. 8.2 and the generalized circuit in Fig. 8.3. Cascode circuits can be used to improve accuracy, but have been omitted for simplicity. The N-port adaptor realization shown in Fig. 8.3 was initially proposed in [4]. Two-port adaptors and programmable three-port adaptors for wave SI filters were proposed in [3] and [7] respectively.

The circuit realizations of N-port parallel and series adaptors are identical – only the placement of γ_k and unity gain mirrors differs. By choosing the mirror ratios to $a_i = \gamma_i$ and $b_i = 1$, a parallel adaptor is realized. In a series adaptor $a_i = 1$ and $b_i = \gamma_i$. The sum A_0 is calculated by adding scaled copies of each input current. A scaled copy of A_0 is then added to an inverse replica of each input current A_k at the corresponding output B_k. The resulting output currents B_k will flow *out from* the parallel adaptors and *into* the series adaptors. According to Fig. 8.1 and the adaptor equations, B_k is defined as flowing out from the adaptors in both cases. This is no practical problem but must be taken care of in a filter implementation.

A 3.3-V CMOS wave SI filter

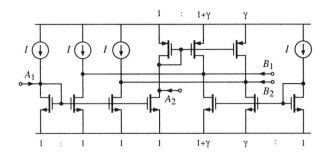

Fig. 8.2 Two-port adaptor realization proposed in [3]. Note that B_1 and B_2 are defined as flowing **into** the adaptor.

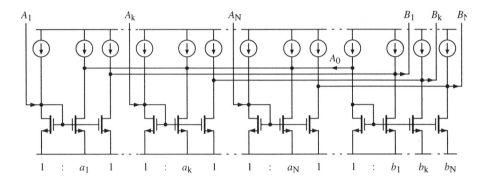

Fig. 8.3 N-port adaptor realization proposed in [4]. Mirror ratios a_i and b_i determine the type of adaptor. Parallel adaptor: $a_i = \gamma_i$ and $b_i = 1$, series adaptor: $a_i = 1$ and $b_i = \gamma_i$. Note that B_k is defined as flowing **out** of the parallel adaptor and **into** the series adaptor.

8.4 Switched-current delay element realization

Delay elements can be realized with switched-current memory cells [8]. A memory cell with reduction of signal-independent (constant) clock feedthrough [9] was used for the WSIF presented in [4]. Its circuit realization is shown in Fig. 8.4 a. For the filter implementation described in this chapter, the SI memory cell shown in Fig. 8.4 b was used due to its much improved linearity. In chapter 4 it was shown that complete cancellation of CFT, and unity gain signal transfer is achieved when

$$\begin{cases} \alpha_4 - \alpha_3 + \alpha_2 - \alpha_1 = 1 \\ \dfrac{1}{\alpha_4} - \dfrac{1}{\alpha_3} + \dfrac{1}{\alpha_2} - \dfrac{1}{\alpha_1} = 0 \end{cases} \tag{8.4}$$

It was also shown that the memory cell can be optimized for power dissipation by choosing $\alpha_1 = 0.3$, $\alpha_2 = 1.4$, which gives $\alpha_3 \approx 0.25$ and $\alpha_4 \approx 0.15$. Unit-size transistors were used in order to improve coefficient matching and therefore the coefficients were rounded to 0.32, 1.4, 0.24 and 0.16 respectively. The simulated current transfer error is shown in Fig. 8.5, where the error of the previously used memory cell has been included for comparison. It is seen that the error magnitude is reduced by approximately 25 times, or ~ 28 dB.

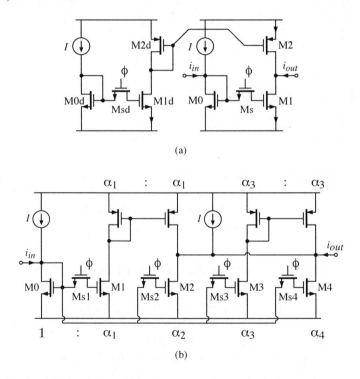

Fig. 8.4 Memory cells. (a) DC-compensation of CFT according to [9], used in [4]. (b) Complete CFT cancellation proposed by the author [10].

A 3.3-V CMOS wave SI filter

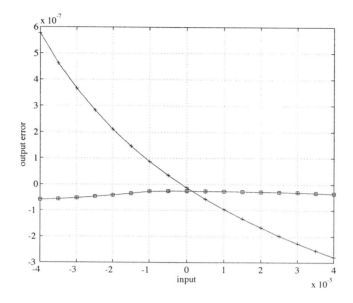

Fig. 8.5 Circuit-level simulated current transfer error vs. input current. (+): Memory cell used in [4] (□): Memory cell proposed in [10].

8.5 Potential for automatic generation

In order to improve accuracy, simple cascode circuits were used in the adaptors and memory elements. The cascode structure together with the nominal transistor sizes are indicated in Fig. 8.6 a. A branch with a gain of α_i is realized by multiplying all MOS transistor widths by α_i. Thus, all node voltages are approximately unaltered, and the circuit design becomes very straightforward. By choosing a proper layout style, one can map the highly regular circuit architecture of the adaptors and the memory elements onto a very regular chip layout. If, for example, all transistors are oriented with their W horizontally and L vertically, as shown in Fig. 8.6 b, each cascode branch will have a fixed height and a width approximately proportional to α_i. This is similar to digital gates used for automated layout and synthesis. It indicates that SI circuits are suitable for automatic layout generation, and possibly even analog synthesis. This can significantly reduce the design time for wave SI filters, as well as for SI circuits in general. In fact, the usefulness of layout automation was illustrated during the design of this particular WSIF. Shortly before submission deadline, the process was updated from 1.0 to 0.8 µm, and important parameters were changed. After a redesign of the nominal transistor dimensions (a unity-gain branch), the layout could be regenerated within a few minutes.

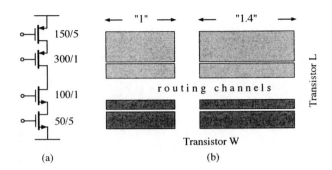

Fig. 8.6 (a) Cascode circuit (b) Layout example of cascode mirror with gain $\alpha_i = 1.4$.

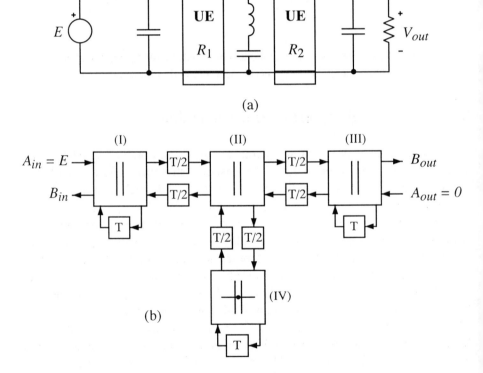

Fig. 8.7 (a) Microwave reference filter. (b) Wave filter signal-flow graph.

8.6 Filter realization

A third order elliptic low-pass filter with 50 kHz bandwidth, 0.01 dB ripple in the pass band and 31 dB attenuation in the stop band was chosen for a test filter realization. The clock frequency is 1 MHz. The microwave reference filter and the corresponding WSIF are shown in Fig. 8.7. The unit elements are inserted so that they don't affect the transfer function, but increase the pipelining and thus improve the circuit settling times. Wave filter structures with inserted unit elements have been treated in detail in [11]. The adaptors and delay elements were realized with the circuits presented in this chapter. Unit-size MOS transistors were used throughout, in order to improve coefficient matching. The nominal "unity-gain" memory or mirror transistor with W/L = 50/5 was split into 25 unit-segments with W/L = 2/5. Therefore every coefficient γ in the adaptors, and α in the delay elements is truncated to a resolution of 0.04, giving the coefficients in Table 8.1.

Coefficient	Value	Truncated value	MOS unit-segments
$\gamma_1^{(I)} = \gamma_3^{(III)}$	1.000	1.00	25
$\gamma_2^{(I)} = \gamma_2^{(III)}$	0.7903	0.80	20
$\gamma_3^{(I)} = \gamma_1^{(III)}$	0.2097	0.20	5
$\gamma_1^{(II)}$	0.1958	0.20	5
$\gamma_2^{(II)}$	1.6084	1.60	40
$\gamma_3^{(II)}$	0.1958	0.20	5
$\gamma^{(IV)}$	0.39	0.40	10
$1+\gamma^{(IV)}$	1.39	1.40	35
Memory cell			
α_1	0.3	0.32	8
α_2	1.4	1.40	35
α_3	0.25	0.24	6
α_4	0.15	0.16	4

Table 8.1 Filter and memory cell coefficient values.

The realization of an N-port adaptor requires N current mirrors, with two outputs each, and one current mirror with N outputs. Knowing from WDF theory [2] that the sum of adaptor coefficients $\gamma_1 + \gamma_2 + ... + \gamma_N \equiv 2$, it is seen from Fig. 8.3 that the total bias current for one adaptor $I_{tot} = 3(N+1)I_{bias}$, where I_{bias} is the bias current in a unity-gain branch. Similarly, the bias currents for the 2-port adaptor and memory cell are shown in Table 8.2, where the total WSIF current has been included. With $I_{bias} = 100$ µA, $V_{DD} = 3.3$ V, and the coefficients from Table 8.1, the estimated power dissipation is 28.5 mW. It should be noted that signal scaling can be used in order to prevent clipping, and to avoid the use of unnecessarily weak signals, internally in the filter. Scaling is described in [2], and in the present application, scaling will only involve the modification of memory cell α-coefficients. The power estimation mentioned above, and the simulation results below, are both for the unscaled WSIF that was actually designed.

Component	Count	Bias current
2-port adaptor	1	$(6+2\gamma)I_{bias}$
3-port adaptor	3	$12 I_{bias}$
Memory cell	12 + 2	$(1+\alpha_1+\alpha_2+\alpha_3+\alpha_4)I_{bias}$
WSIF total		$(56+2\gamma+14[\alpha_1+\alpha_2+\alpha_3+\alpha_4])I_{bias}$

Table 8.2 Bias current.

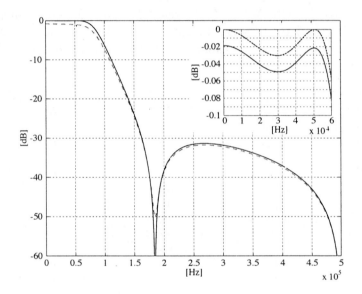

Fig. 8.8 Simulated frequency response. Present design (solid), with DC-compensation only (dashed), and ideal passband response (dotted).

8.7 Simulation results[2]

The filter circuit was simulated with Cadence Spectre™, using level 2 MOS transistor models for a 0.8 μm standard CMOS process, and the filter coefficients according to Table 8.1. A 3.3 V power supply was used. Fig. 8.8

---

[2] The simulations presented in this chapter are re-simulations of the presented chip design, using Cadence Spectre™ models for the same process. HSPICE™ was used for the simulations presented in the original publication [4], and during the design of the present circuit which was first presented in [12].

shows the simulated frequency response. The level shift of approximately 0.02 dB and the slight distortion of the frequency response are due to current transfer errors in the memory cells and adaptors. The main error sources are clock feedthrough and current mirror errors.

8.8 Experimental results

A wave SI filter test chip, designed in a 0.8 μm CMOS process was implemented. A micro-photograph of the filter layout is shown in Fig. 8.9. The filter occupies about 3.3 mm² of active area, including the necessary input and output circuitry. The adaptors are seen in the middle, with delay elements on both sides. Due to fabrication errors, the circuit unfortunately malfunctioned, and therefore no measurement data can be presented.

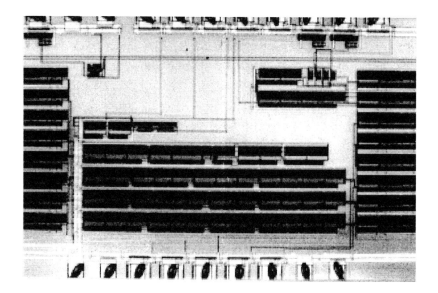

Fig. 8.9 Wave SI filter test chip microphotograph.

8.9 Summary

In this chapter, we have seen the use of wave SI filters for analog sample-data signal processing. Current-mode realizations of N-port adaptors were given. With these circuits available, the number of possible WSIF implementations is essentially increased. The filters are easy and straightforward to design, and many of the well-established WDF design methods can be used. They rely on a few building blocks, i.e. the delay and the adaptors. Since the general topology is fixed and adaptor coefficients are in fact only determined by MOS transistor ratios, there is also a possibility for automated generation of such filters. The circuit architecture is highly regular, and the possibility for automatic generation of SI filters may increase their significance in commercial applications. Such auto-generation of layout was used for this design. An elliptic wave SI filter was laid out in a 0.8 μm double-metal CMOS process. It occupies about 3.3 mm^2 active chip area. Simulation results indicate that the circuit works well at 3.3 V power supply voltage, with a straightforward cascode approach. By using special low-voltage design techniques it should be possible to operate at much lower voltages.

REFERENCES

[1] J. B. Hughes, N. C. Bird, and I. C. Macbeth, "Switched Currents – A New Technique for Analog Sampled-Data Signal Processing", *Proc. of Int. Symp. Circuits And Systems (ISCAS)*, Portland, Oregon, pp. 1584-1587, May 1989, IEEE.

[2] L. Wanhammar, *DSP Integrated Circuits*, Academic Press Series in Engineering, 1999.

[3] A. Rueda, A. Yúfera, J. L. Huertas, "Wave analogue filters using switched-current techniques", *Electron. Lett.*, 1991, Vol. 27, No. 16, pp. 1482-1483, Aug. 1991.

[4] B. Jonsson, and S. Eriksson, "Current-Mode N-port Adaptors for Wave SI Filters", *Electron. Lett.*, Vol. 29, No. 10, pp. 925-926, May 1993.

[5] A. Fettweis, "Digital Filter Structures Related to Classical Filter Networks", *Arch. Elektr. Übertragungstech.*, Vol. 25, No. 2, pp. 79-89, Feb. 1971.

[6] C. Toumazou, F. J. Lidgey, D. G. Haigh, (Eds.) *Analogue IC design: the current-mode approach*, IEE Circuits and Systems series 2, 1990.

[7] A. Yúfera, A. Rueda, J. L. Huertas, "A Methodology for Programmable Switched-Current Filters Design", *Proc. of European Conf. Circuit Theory and Design*, Davos, Switzerland, pp. 317-322, Aug. 1993, Elsevier.

[8] C. Toumazou, J. B. Hughes, N. C. Battersby, (Eds.) *SWITCHED-CURRENTS an analogue technique for digital technology*, IEE Circuits and Systems series 5, 1993.

[9] H. C. Yang, T. S. Fiez, and D. J. Allstot, "Current-Feedthrough Effects and Cancellation Techniques in Switched-Current Circuits", *Proc. of Int. Symp. Circuits And Systems (ISCAS)*, pp. 3186-3188, May 1990, New Orleans, IEEE.

[10] B. Jonsson, and S. Eriksson, "New Clock-Feedthrough Compensation Scheme for Switched-Current Circuits", *Electron. Lett.*, Vol. 29, No. 16, pp. 1446-1447, Aug. 1993.

[11] U. Ullrich, *Effektive Verfahren zur Ermittlung des Signalrauschverhältnisses von Digitalfiltern unterschiedlischer Struktur*, Dissertation Ruhr-University Bochum, Germany, 1976

[12] B. Jonsson, and S. Eriksson, "A Low Voltage Wave SI Filter Implementation Using Improved Delay Elements", *Proc. of Int. Symp. Circuits And Systems (ISCAS)*, London, UK, pp. 5.305-5.308, May 1994, IEEE.

Chapter 9

A 3.3-V CMOS Switched-Current Delta-Sigma Modulator[1]

This chapter presents the design of a second-order switched-current delta-sigma modulator having equal first and second integrator output swing. It was implemented using the clock-feedthrough compensated first-generation switched-current circuits described in chapter 4. Experimental results show that the modulator has a small chip area, low supply voltage, and low power dissipation. The measured dynamic range is approximately 11 bits.

9.1 Introduction

Delta-sigma (Δ-Σ) modulation has been employed in oversampling A/D converters. It has mainly been realized using the *switched-capacitor* (SC) technique [1]. The *switched-current* (SI) technique, on the other hand, offers some interesting advantages over the SC technique, as pointed out in [2] and in chapter 1. The design of Δ-Σ data-converters in general is thoroughly described in [3], while switched-current Δ-Σ A/D-converters are extensively treated in [4]. Second-order Δ-Σ modulators were proposed in [5-8]. Since the integrators had no delay, the settling of the modulator in [6] was poor. Cascading two first-order stages to realize a second-order modulator, as was done in [7], has a serious limitation in performance due to the mismatch between analog and digital circuits as pointed out in [1]. The modulators proposed in [5] and [8] resembled the SC modulator in [9], which has

---

[1] N. Tan, B. Jonsson, and S. Eriksson, "3.3 V 11 Bit Delta-Sigma Modulator Using First-Generation SI Circuits", *Electron. Lett.*, Vol. 30, No. 22, pp. 1819-1821, Oct. 1994.

documented the most favorable performance. However, the signal swings at the two integrator outputs had different range. Therefore, concerning signal range, they were not optimal.

9.2 Modulator structure

The second-order $\Delta\text{-}\Sigma$ modulator structure is shown in Fig. 9.1. Both integrators have one sample delay. The input signal i_{in} is a current, and the output d_{out} is a voltage having CMOS logic levels. The comparator produces the 1-bit signal by sensing the direction of current flow from the second integrator. To feedback the quantized signal, d_{out} is used as an input to a pair of 1-bit D/A converters, one feeding the first integrator and the other feeding the second integrator.

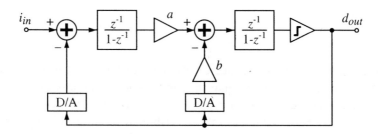

Fig. 9.1 Modulator structure.

In an SC integrator, a scaling factor is directly realized by changing the ratio of the integration capacitor and the sampling capacitor. However, in an SI integrator, integration and scaling are done separately. Scaling can either precede or follow integration. A scaling factor before the first integrator only scales the input signal, and the scaling before the second integrator is realized by changing parameters a and b. Due to the stability requirement, b must be

$$b = 2a \tag{9.1}$$

and the values of a and b have an influence on the signal swing at the second integrator output. The factor c does not have any influence on the functionality because the current comparator only detects the current direction. Optimal selection of $\{a, b, c\}$ is discussed in the following section.

9.3 System level simulations

In Fig. 9.2, the simulated histogram of the integrator outputs with $a = 0.5$ and $b = 1$ is shown. The input is a 1-kHz sinusoidal with an amplitude equal to half of the feedback current value, and the clock frequency is 1.024 MHz. It is seen that both integrators have the same range of signal swing. The modulator (with $a = 0.5$ and $b = 1$) only requires a signal range in both integrators slightly larger than twice the full-scale input range, i.e., twice the output current value of the D/A converters.

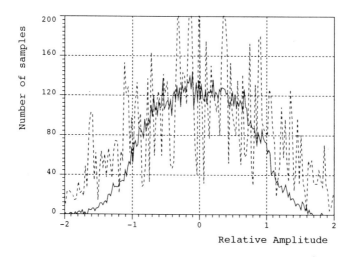

Fig. 9.2 Histogram of the integrator outputs: First integrator (dashed), second integrator (solid).

If $a = 0.25$ and $b = 0.5$ as in [8], the signal swing of the second integrator output will be half that of the first integrator. Thus the signal range of the second integrator is not fully utilized. If $a = 1$ and $b = 2$, the signal swing of the second integrator will be twice as large as that of the first integrator. The second integrator will call for a *larger* signal range, leading to unnecessary power dissipation. Therefore, the modulator with $a = 0.5$ and $b = 1$ gives the best performance. The choice of the scaling coefficient c has no effect on the signal swing within the integrators. It only has slight influence on the design of the current comparator and on the output current mirror of the second integrator. Here, $c = 1$ was chosen.

9.4 Circuit implementation

The core building block in this design is the memory cell used to implement the integrators. The performance of a $\Delta\text{-}\Sigma$ modulator is often limited by the memory cells used in the first integrator. Since the second-generation memory cell *ideally* has no mismatch error it should have an advantage over the first-generation memory cell. This advantage is, however, reduced in many practical SI circuits, because current mirrors are always needed to generate multiple current outputs. There is also the adverse effect of the transient current glitches in second-generation SI circuits [10], and a gain error caused by the linear clock-feedthrough (CFT) term, as shown in [11]. Therefore, contrary to earlier SI designs [5-8], first-generation SI circuits were used in this design. A first-generation memory cell with a CFT compensation scheme according to [12] is shown in Fig. 9.3. Under the conditions given in chapter 4, both signal-independent and signal-dependent CFT errors are canceled. Cascode transistors (not shown) are used in order to increase the input-output conductance ratio.

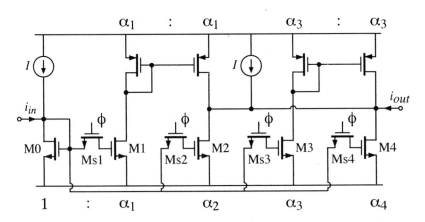

Fig. 9.3 First-generation memory cell with CFT compensation.

Each integrator is realized as shown in Fig. 9.4, using a cascade of two memory cells for the delay, and a set of current mirrors to perform the current duplication and scaling. The summation at the input node is done by simply wiring the two current signals together. A CMOS inverter is used as a current comparator, according to [6]. The 1-bit D/A converters are current sources that are switched according to the output of the current comparator.

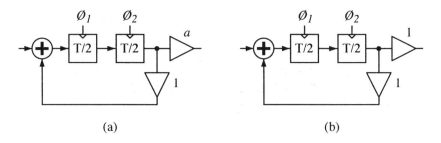

Fig. 9.4 Integrator realizations using SI memory cells: (a) First integrator including output scaling. (b) Second integrator.

9.5 Experimental results

The whole modulator was designed and implemented in a 0.8 μm digital CMOS process. It occupied an active area of 0.48 mm². A microphotograph of the implemented chip is shown in Fig. 9.5. A single 3.3 V power supply was used, and the measured power consumption was about 6.6 mW. The measured[2] *signal-to-noise-and-distortion ratio* (SNDR) vs. input level is shown in Fig. 9.4. The input is a 2 kHz sinusoidal current, the clock frequency is 2.45 MHz, and the *oversampling ratio* (OSR) is 128. The *dynamic range* is seen to be 67 dB, while the *peak SNDR* is limited by harmonic distortion to 60 dB (9.7-b). It is believed that the performance drop seen in Fig. 9.6 is due to CFT effects in the memory cell. As discussed in [11], and chapter 4, the memory cell can be optimized for different goal functions – power, area, distortion, etc. In this design, memory cells with near-optimum power dissipation were used. In such memory cells, one of the α-coefficients become rather small (here $\alpha_4 \approx 0.15$), which means that transistor M4 in Fig. 9.5 should be small. The charge injected by CFT will then cause a much larger gate voltage shift on M4 which, depending on the operating point, may approach cut-off for large negative input currents. In such a case, the conditions for complete cancellation of CFT are no longer valid, and therefore distortion will increase significantly. This problem can be avoided by selecting a less aggressive power optimization of the memory cells, as was indicated in [11].

---

[2] The measurements were done by Mr. T. Ritoniemi of *VLSI Solution Oy*, Finland.

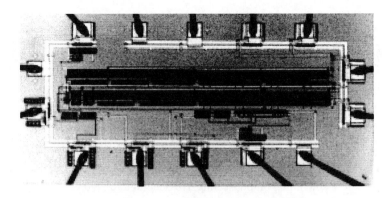

Fig. 9.5 Chip micro-photograph.

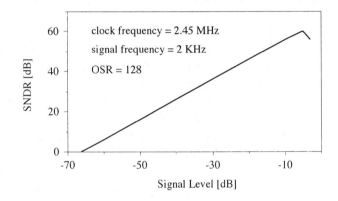

Fig. 9.6 The measured SNDR versus input current. The 0-dB level is 20 μA.

Parameter	f_{in} = 2 kHz, f_s = 2.45 MHz
SNDR [dB]	60
DR [dB]	67
ENOB	9.7
Common parameters	
B/W$_{in}$ [kHz]	0 – 9.6
OSR	128
VDD [V]	3.3
P [mW]	6.6
A [mm^2]	0.48

Table 9.1 Summary of experimental results.

9.6 Summary

In this chapter we have discussed the design and implementation of second-order SI Δ-Σ modulators. A second-order Δ-Σ modulator design with equal first and second integrator output swing was presented. The equal-swing modulator was implemented in a 0.8 μm digital CMOS process, using clock-feedthrough compensated first-generation SI memory cells described in chapter 4. It occupied an active area of 0.48 mm^2 and dissipated 6.6 mW from a single 3.3 V power supply. The measured dynamic range was about 11 bits with a decimation ratio of 128. The competitive performance, when compared to earlier[3] implementations, is due to the modulator structure and the clock feedthrough compensated SI memory cells[4].

REFERENCES

[1] J. C. Candy, and G. C. Temes, (Eds.) *Oversampling Delta-Sigma Data Converters: Theory, Design and Simulation*, IEEE Press: New York, 1992.

[2] C. Toumazou, J. B. Hughes, N. C. Battersby, (Eds.) *Switched-Currents: an analogue technique for digital technology*, IEE Circuits and Systems series 5, 1993.

[3] S. R. Norsworthy, R. Schreier, and G. C. Temes (Eds.): *Delta-Sigma Data Converters: Theory, Design and Simulation*, IEEE Press, 1997.

[4] N. Tan, *Switched-Current Design and Implementation of Oversampling A/D Converters*, Kluwer, 1997.

[5] S. J. Daubert, and D. Vallancourt, "A Transistor-Only Current-Mode $\Sigma\Delta$ Modulator", *IEEE J. Solid-State Circuits*, Vol. 27, No. 5, pp. 821-830, May 1992.

[6] P. J. Crawley, and G. W. Roberts, "Switched-Current Sigma-Delta Modulation for A/D Conversion", in *Proc. of Int. Symp. Circ. Syst.*, San Diego, California, May 1992, pp. 1320-1323.

[7] C. Toumazou, and G. Saether, "Switched-current circuits and systems", in *Proc. of Int. Symp. Circ. Syst.*, London, UK, May 1994, Vol. Tutorials, pp. 459-486.

[8] N. Tan, and S. Eriksson, "A Fully Differential Switched-Current Delta-Sigma Modulator Using a Single 3.3-V Power-Supply Voltage", in *Proc. of Int. Symp. Circ. Syst.*, London, UK, May 1994, pp. 5.485-5.488.

[9] B. E. Boser, and B. A. Wooley, "The design of sigma-delta modulation analog-to-digital converters", *IEEE J. Solid-State Circuits*, vol. 23, pp. 1298-1308, Dec. 1988.

[10] P. M. Sinn, and G. W. Roberts, "A comparison of first and second generation switched-current cells", in *Proc. of Int. Symp. Circ. Syst.*, London, UK, May 1994, Vol. 5, pp. 301-304.

---

[3] This work was originally presented in 1994. Since then considerably higher performance has been reported, e.g., for the S^{2}I-based Δ-Σ modulators in [13], which have the highest performance reported for any SI implementation: 14-b linearity and SNR > 80 dB.

[4] In a comparison between eight different SI Δ-Σ modulator implementations using different memory cells [14], it was found that the present design also had the highest resolution.

[11] B. Jonsson, and N. Tan, "Clock-Feedthrough Compensated First-Generation SI Circuits and Systems", *Analog Integrated Circuits and Signal Processing*, Vol. 12, No. 4, pp. 201-210, Apr. 1997.

[12] B. Jonsson, and S. Eriksson, "New Clock-Feedthrough Compensation Scheme for Switched-Current Circuits", *Electron. Lett.*, Vol. 29, No. 16, pp. 1446-1447, Aug. 1993.

[13] N. Moeneclaey, and A. Kaiser, "Design Techniques for High-Resolution Current-Mode Sigma-Delta Modulators", *IEEE J. Solid-State Circ.*, Vol. 32, No. 7, pp. 953-958, July 1997.

[14] N. Tan, "Switched-Current Delta-Sigma A/D Converters", *Analog Integrated Circuits and Signal Processing*, Vol. 9, No. 1, pp. 7-24, Jan. 1996.

Chapter 10

A 3-V Wideband CMOS Switched-Current A/D-Converter[1]

The simulated and measured performance of an experimental wideband CMOS A/D converter design is presented in this chapter. Fully-differential first-generation switched-current circuits with common-mode feedforward were used to implement a 1.5-b/stage pipelined architecture in order to evaluate the switched-current technique for digital radio applications. With f_{in} = 1.83 MHz, the measured SFDR = 60.3 dB and SNDR = 46.5 dB at f_s = 3 MHz. Although this 3 V design was fabricated in a standard digital 5 V, 0.8 μm CMOS process, a high bandwidth was achieved. Since the ADC maintains an SNDR ≥ 40 dB for input frequencies of more than 20 MHz, it has the highest input bandwidth reported for any CMOS switched-current A/D-converter implementation. Its sample rate can be increased by parallel, time-interleaved, operation. Measurement results are compared with the measured performance of other wideband switched-current A/D converters and found to be competitive also with respect to area and power efficiency.

10.1 Introduction

A wide range of analog interfacing circuits are needed to meet the needs stated by the evolution of digital signal processing technology. Design specifications span from highest possible speed and accuracy to ultra-low power supply voltage, low power dissipation or lowest total manufacturing

---

[1] B. E. Jonsson, and H. Tenhunen, "A 3V Wideband CMOS Switched-Current A/D-Converter Suitable for Time-Interleaved Operation", To be published in *Kluwer J. Analog Integrated Circuits and Signal Processing*, June 2000.

cost. While the *switched-capacitor* (SC) technique has become the main tool for implementing CMOS A/D converters, there is a need to explore alternative solutions in some applications. SC circuits require linear capacitors, and their signal swing is directly limited by the supply voltage. This is a drawback when designing *one-chip*, mixed analog-digital systems, since a high power supply voltage is incompatible with state-of-the-art digital CMOS technology. In addition to that, the extra processing steps required for capacitors will add to the manufacturing cost. *Switched-current* (SI) circuits, on the other hand, are compatible with digital VLSI technology and therefore suitable for integration with digital CMOS circuits. SI circuits use the MOS transistor gate parasitic capacitance to memorize the gate voltage and thus sample the drain current. The need for accurate and linear capacitors is thereby eliminated. Since the signals are represented by currents rather than voltages, the signal swing is only indirectly limited by the supply voltage [1].

Switched-current A/D converters were proposed in [2] and [3]. Up to 13-b linearity has been demonstrated [4], although only at a 5.7 kHz sample rate. In spite of the potential for high-speed [5] and low-voltage [6] operation, only a few CMOS SI ADC implementations with **measured** results at $f_{in} \geq 1$ MHz has been reported [7-10]. Many "high-speed" SI ADCs in the literature have only been characterized with low-frequency input signals. As a result, their performance is comparable to narrow-band Δ-Σ A/D converters [11]. Of the wideband implementations, only one design [10] is for 3 V power supply, and special transistors with $V_T = 0.4$ V were used for the sampling switches in order to increase f_s. In many emerging radio systems applications, wideband ADCs with sampling rates from a few megahertz up to 50 MHz are of special interest. In this chapter we focus on CMOS SI techniques with potential for such performance.

The ADC presented in this chapter is a 3 V design implemented in a standard 5 V CMOS process having a typical V_T of 0.8 V. It has been characterized with input signals from 300 kHz to 100 MHz, thereby showing its static linearity as well as its wideband performance. A 1.5-b/stage pipeline architecture was chosen to evaluate the switched-current technique for wideband data-converter applications. In sections 10.3 and 10.4, a brief description of the 1.5-b/stage architecture and its circuit implementation is given. System and circuit-level simulations are included in section 10.5, and measurement results presented in section 10.6. The performance is compared with previously reported implementations in section 10.7, where the measured *effective number-of-bits* (ENOB) vs. f_{in} and f_s is plotted for all designs if available. Furthermore, the *area* and *power efficiency* related to ENOB and input bandwidth is examined. First, a review of the evolution of

CMOS SI ADCs, with emphasis on implemented and measured designs is given in the following section.

10.2 Switched-current A/D converters

The early work by Nairn and Salama [12-13] appeared before the switched-current memory cell was proposed [14]. It is not really *switched-current* but rather *current-mode* ADCs where algorithmic A/D conversion is performed by a cascade of N bit-cells. A true switched-current approach was used both by Nairn et. al. [2, 15-16] and Robert et. al. [3]. Dynamic current memories, later known as 2nd generation SI memory cells, were used to implement a gain of two that is not depending on transistor matching, and thus potentially mismatch free. Deval et. al. presented an alternative solution where the reference current is dynamically divided by two while the input signal is merely replicated from cell to cell [17]. The divide-by-two operation thus operates on DC currents and can be optimized for accuracy rather than speed. A similar algorithm was used in [18] although with a static scaling of I_{ref}. In addition, various SI delta-sigma converters have been presented, mainly for audio applications [11].

To the best of the author's knowledge, the highest SI ADC resolution measured to this day is reported in [4], where 13-bit linearity (INL ≈ +/- 1 LSB @ 14-b) was measured at f_s = 5.7 kHz for a cyclic implementation. In many signal processing applications it is more important to know the spectral characteristics rather than the static linearity. Kim et. al. were first to report a measured output *spectrum* [19], which shows 2nd and 3rd harmonics to be 48 and 69 dBc respectively at a 120 kS/s sample rate. This implicates an upper limit on the effective number-of-bits of 7.7 – if measured noise is negligible. A *redundant signed digit* (RSD), or 1.5-b/stage, pipelined ADC was implemented by Macq and Jespers [20]. Measured INL is 0.8 LSB at the 10-b level while sampling at 550 kS/s. The RSD algorithm has later been used by Sugimoto et. al. [10] and the author [21]. This will be described in section 10.3. An algorithm with similar properties regarding offset errors was also used in [22]. Digital correction is allowed for by overlapping signal ranges in the first bitcells similar to what is common in multibit pipelined ADCs. Dynamic multiplication by two and S^2I circuits were used to realize a high accuracy 1.5 V A/D converter. Experimental results are found in [6]. Switched-current A/D converters measured with high sampling rate **and** input frequencies ≥ 1 MHz were reported by Wu et. al. [7] and Bracey et. al. [8-9, 23]. A 3 V SI ADC with comparable performance to above was published in [10]. In section 10.7, its measured performance is compared with the low-voltage SI ADC presented in this chapter.

10.3   RSD A/D converter architecture

Several parameters are critical for the overall performance of an ADC. On the system level, interstage gain accuracy and low comparator offset are essential in typical ADCs. The *redundant signed digit (RSD)*, or 1.5-b/stage, algorithm was chosen for this implementation because of its insensitivity to comparator offsets, and its reduced sensitivity to interstage gain errors [24]. Because simple comparators can be used, it improves area and power performance. In the case that the ADC accuracy is limited by MOS transistor matching in the interstage amplifier, it is also improved by a factor of 2, i.e., by one bit. The RSD algorithm for N bits resolution is described by the following equations:

$$I_i(1) = i_{in}$$
$$I_o(j) = 2I_i(j) + q_j I_{ref} - p_j I_{ref} \quad j = 1 \ldots N-1$$
$$p_j = 1 \quad \text{if} \quad I_i(j) \geq I_P \quad \text{else} \quad p_j = 0$$
$$q_j = 1 \quad \text{if} \quad I_i(j) \leq I_Q \quad \text{else} \quad q_j = 0 \tag{10.1}$$

$$d = \sum_{j=1}^{N-1} (p_j - q_j) 2^{N-1-j} \tag{10.2}$$

A pipelined implementation was chosen in order to increase sampling rate. The comparator levels I_P (I_Q) can be selected between $0 \ldots + I_{ref}/2$ ($0 \ldots - I_{ref}/2$) and does not have to be equal in magnitude. If chosen to $+(-)I_{ref}/4$, the converter has optimum tolerance to offset errors [24]. Static as well as temporary offsets (noise & glitches) with a magnitude up to $I_{ref}/4$ can occur without loss of converter accuracy. A block level schematic of the pipelined ADC is shown in Fig. 10.1. Unless the signal to be digitized is already a current, a V/I converter is needed at the input. The analog input is converted sequentially into the RSD codes p and q. After serial-to-parallel conversion, the binary code representation d of the input sample is calculated according to Eq. 10.2. Unlike a typical binary ADC, each RSD converter "bitcell" performs a three-level decision $\{-1, 0, +1\}$ rather than the more common two-level decision $\{-1, +1\}$. The redundant information extracted at each bit-level allows for the digital correction. From Eq. 10.2 it is clear that the range of d is $\{+/- 1111\ldots11\}$, and therefore N bit resolution is achieved with $N-1$ bitcells. The RSD conversion algorithm is treated in detail in [24]. In short, it can be described as taking a bit-level decision, $p = 1$ ($q = 1$), only if the input is *significantly* above (below) zero. If not, $p = q = 0$, and the signal is scaled by two and passed to the next bitcell.

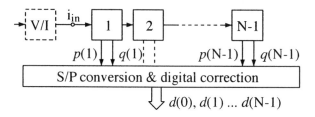

Fig. 10.1 SI ADC block diagram showing the V/I converter, N-1 bitcells and the digital correction.

10.4 Circuit implementation

The A/D converter was implemented in a 5 V, 0.8 μm standard digital CMOS process (*AMS cyb*). Current mirrors and first-generation SI memory cells [25] were used to realize the scaling and sampling operations. The three main performance limitations associated with SI circuits are clock-feedthrough (CFT), conductance ratio errors, and noise. A fully-differential approach was chosen in order to suppress external noise sources, such as substrate noise [26], and a fully-differential *CFT compensated* S/H with a gain of two was developed as shown in Fig. 10.2 a. The CFT compensation scheme is that proposed by Fiez et. al. [27]. By using this technique, signal dependent CFT errors are canceled, leaving only a constant offset term. This offset, as well as external noise will appear as a common-mode signal, and therefore *common-mode feedforward* (CMFF) cancellation [28] was used to ensure a proper operating region in all bitcells. The common-mode signal is relatively small and therefore CMFF is not needed throughout the pipeline. However, for simplicity, all bitcells were made identical. Therefore, area and power dissipation are sub-optimal in this experimental implementation. Cascodes (not shown) were used to increase the output conductance and thereby reduce input/output conductance ratio errors. Interdigitized layout style, rather than common-centroid was used to ensure a dense layout with good matching. The power supply wires were carefully designed in order to reduce current mismatch due to resistive voltage drops as shown in [29]. A layout plot of the ADC pipeline is shown in Fig. 10.3 with digital correction excluded.

The current comparator proposed by Träff [30] was used to determine the p and q values of each bitcell. Since the requirements on the comparator are greatly relaxed by the RSD algorithm, a simple implementation using minsize transistors was sufficient. A 1.5-b D/A converter was realized with a PMOS current source, an NMOS current sink and current-steering switches controlled by p and q. When $p = 1$ or $q = 1$, one of the sources is connected to the positive signal and the other to the negative signal. Therefore, the level-shift at I_Q and I_P shown in Fig. 10.4 b will always be $|I_{sink}| + |I_{source}|$ in the differential signal. Accurate matching between the sink and the source is not needed, and the effective reference level becomes

$$I_{ref,eff} = |I_{source}| + |I_{sink}| \tag{10.3}$$

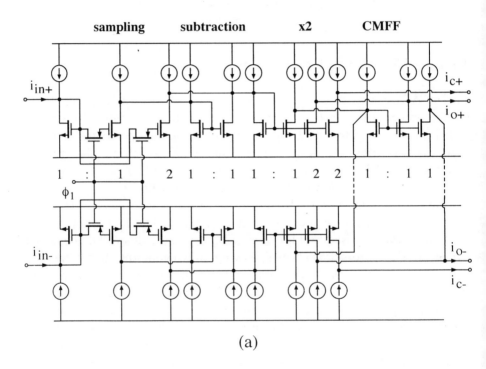

(a)

Fig. 10.2 SI bitcell realization: (a) Current sample-and-hold with common-mode feedforward.

A 3-V wideband CMOS switched-current A/D-converter

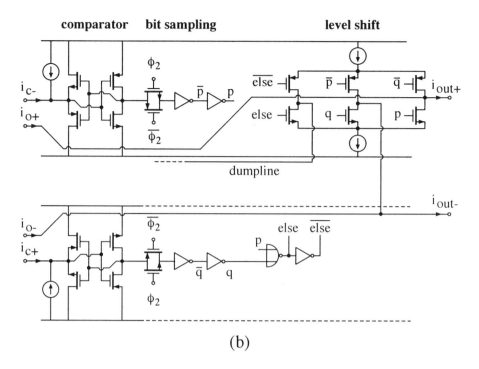

Fig. 10.2 contd. SI bitcell realization: (b) Current comparators and 1.5-b DAC.

Fig. 10.3 Layout of switched-current ADC pipeline with digital correction excluded. The layout of each bitcell is similar to the topology of the circuit schematics in Fig. 10.2. Samples propagate from left to right.

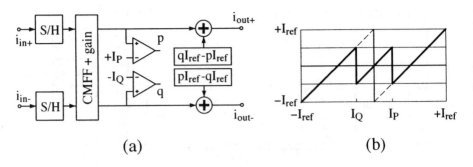

Fig. 10.4 Bitcell. (a) Block level schematics. (b) Transfer characteristics: 1.5-b (solid) and 1-b per stage (dashed).

10.5 Simulation results

The design has been carefully tested with circuit and system level simulations in order to predict the performance. System level simulations display the effects of interstage gain errors, offset and component mismatch while circuit level simulations capture settling errors and non-linear effects such as CFT and the current-mirror transfer function. Analyzing the *spectral* performance by circuit level simulation has the following important advantages:
- The frequency-domain behavior is predicted with realistic input signals using the most accurate circuit model available.
- It is possible to directly compare simulation and measurement results.
- Design errors can be detected at an early stage.

Because noise sources such as thermal noise and substrate switching noise are usually not accounted for, simulations can only predict an *upper limit* of the performance. The measured noise-level may be significantly higher than the simulated, depending on the circuit design. Device mismatch effects are also difficult or time-consuming to include.

10.5.1 System level simulations

Consider the 1.5-b algorithm described by Eqs. 10.1 and 10.2 and the circuit and block level schematics in Figs. 2 and 4. Several non-ideal effects are included by rewriting Eq. 10.1 as

$$I_o(j) = 2(1+\varepsilon_j)I_i(j) + \Delta_{CFT} +$$
$$+ q_j(I_{ref} + \delta_{rpj}) - p_j(I_{ref} + \delta_{rnj})$$
$$p_j = 1 \quad \text{if} \quad I_i(j) \geq I_P + \Delta_P \quad \text{else} \quad p_j = 0$$
$$q_j = 1 \quad \text{if} \quad I_i(j) \leq I_Q + \Delta_Q \quad \text{else} \quad q_j = 0 \tag{10.4}$$

The *interstage gain error*, ε_j, represents the MOS transistor mismatch and the signal loss due to finite conductance ratios as the signal is transferred between internal current mirrors. Although the chosen current S/H cancels signal-dependent CFT, a small constant *CFT residue*, Δ_{CFT}, remains. Offsets in the reference current source and sink are represented by δ_{rpj} and δ_{rnj} and comparator offsets by Δ_{Pj} and Δ_{Qj}. Total matching of two drain currents depends on geometric mismatch as well as device parameter variation. The MATLAB™ Monte Carlo simulations were performed with the following statistical model: If the matching between each MOS transistor pair is σ_{MOS}, it can be shown that the total gain, G, through the current S/H and the multiply-by-two circuit in Fig. 10.2 a has

$$\sigma_G = \left(\sqrt{14}/2\right)\sigma_{MOS} \approx 1.87\sigma_{MOS} \tag{10.5}$$

With careful layout design, it is reasonable to assume that close-by MOS transistors match to approximately 1%. Variance for ε was therefore set to 1.87% and 1% for δ and $\Delta_{P,Q}$. The gain errors were centered around -0.1% representing the loss due to conductance ratio errors. A Δ_{CFT} of 50 nA was used, in accordance with circuit level simulations. The performance is mainly limited by the interstage gain matching since the RSD algorithm is insensitive to offset errors [24]. The resulting *spurious-free dynamic range* (SFDR) distribution is shown in Fig. 10.5. The mean SFDR value is approximately 55 dB which is *lower* than the measured peak SFDR of 62.9 dB presented in section 10.6.

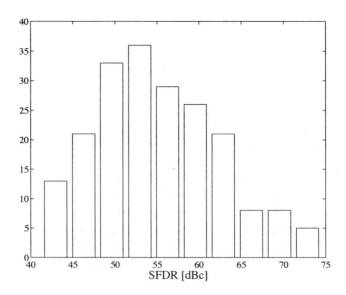

Fig. 10.5 Expected SFDR yield with 1% MOST matching and 0.1% conductance ratio loss.

10.5.2 Circuit level simulations

Simulation of spectral performance was possible due to the relative simplicity of SI circuits as compared to their switched-capacitor (SC) counterparts. The entire ADC core was simulated using *Cadence SPECTRE*™ analog circuit simulator in "conservative" accuracy mode. One 2048 point FFT required data from approximately 20 hours of simulation. The RSD digital outputs were converted to numerical data and analyzed in MATLAB™ using a Kaiser window. The simulated FFT of a *full scale – 4 LSB*, 15.85 MHz sinusoidal current converted at 6.4 MS/s is shown in Fig. 10.6. It is seen that SFDR is limited by the third harmonic to 47.9 dB. *Signal to noise and distortion ratio* (SNDR) is 46.2 dB meaning that the simulated ENOB is 7.3 even for input frequencies in the *fifth* Nyquist band. It shows the high input bandwidth of the sample-and-hold stage. The ADC was characterized at f_s = 6.4 MHz for different f_{in}, and the results are shown in Fig. 10.7. Simulated SNDR is larger than 50 dB (8 bits) up to more than 6 MHz, and peak ENOB is 9.6. Although the circuit-level simulations overestimate the noise performance they seem to capture quite well the distortion-limited performance degradation at higher frequencies. It should be noted that the distinct performance drop above 3 MHz is also found in the corresponding measurements at 2 MHz.

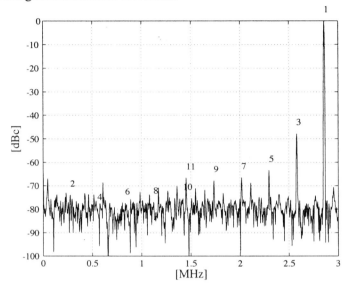

Fig. 10.6 A/D converter output spectrum from circuit-level simulation with $f_s = 6.4$ MHz and $f_{in} = 15.85$ MHz.

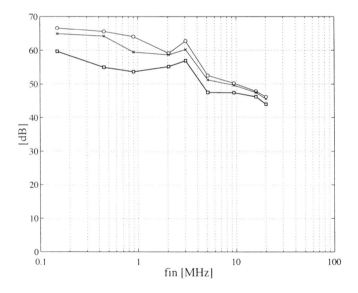

Fig. 10.7 Simulated spectral performance vs. input frequency with $f_s = 6.4$ MHz. Legend: (o) SFDR, (x) THD, and (□) SNDR.

Parameter	f_{in} = 150 kHz	f_{in} = 3.05 MHz	f_{in} = 15.85 MHz
SFDR [dB]	66.6	62.7	47.8
SNDR [dB]	59.6	57.0	46.2
ENOB [bits]	9.6	9.1	7.3
	Common parameters		
f_s [MHz]	6.4		
VDD [V]	3		
P [mW]	90		
Area [mm^2]	2.7		

Table 10.1 Summary of ADC circuit level simulations.

10.6 Experimental results

Extensive testing of linearity and spectral performance has been done. A simple fully differential off-chip V/I converter consisting of a transformer and two precision resistors was used to feed the ADC with current-domain input signals during test. Simulations indicated that, in spite of the non-linear interaction with the current-mirror input, this simple V/I converter could achieve at least 50-55 dB SFDR. This was also confirmed by the measurements. Converter linearity was measured using histogram methods [31-32]. Figure 10.8 shows the differential and integral non-linearity (DNL and INL) estimated from 131 kS of a 1.13 kHz sinusoidal input sampled at 3 MHz. Peak DNL and INL is 0.9 and 4.9 LSB respectively. The distinct x^3 shape of the INL reflects the third-order harmonic distortion of the input stages more than interstage gain or offset errors. A similar shaping of the INL curve is seen in MATLAB™ simulations of the ADC when a third-order harmonic is added to the input signal. There are no missing codes or decision levels. The 2048 pt. FFT of a 1.83 MHz, -1 dBFS, sinusoidal current digitized at 3 MS/s is shown in Fig. 10.9. The first 11 harmonics are indicated. It is seen that SFDR is limited by the third order harmonic to 60.3 dB. THD = -52.2 dB and SNDR is limited by noise to 46.5 dB (7.43-b). Spectral performance vs. input frequency at 3 MS/s is shown in Fig. 10.10. Signal quality is optimal for input frequencies up to 2 MHz where a slight degradation is observed, as predicted by the circuit level simulations. High-frequency performance is limited by the track-mode non-linearity of the input track-and-hold, but not until $f_{in} \geq 25$ MHz. The effective resolution has dropped 1-b from its low-frequency value at $f_{in} = 25$ MHz, where ENOB is still 6.2. This demonstrates that the ADC has a high input bandwidth and works well even with undersampling. Therefore it can be used in a parallel, time-interleaved, high-speed ADC architecture, as described in chapter 11.

A 3-V wideband CMOS switched-current A/D-converter

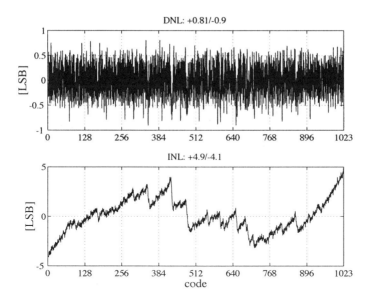

Fig. 10.8 DNL and INL @ 10-b with f_s = 3 MHz and f_{in} = 1.13 MHz.

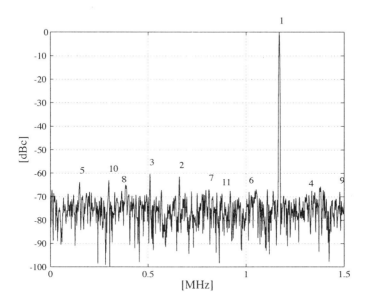

Fig. 10.9 FFT of a 1.83 MHz, -1 dBFS, sinusoidal current digitized at 3 MS/s.

The ADC was also characterized with f_{in} = 20.13 MHz for various sample rates, and the result is shown in Fig. 10.11. Distortion starts to increase significantly and limits the performance for sample rates above 4 MS/s due to bitcell internal settling errors. This is a lower sampling rate than predicted by the circuit-level simulations where the effects of layout parasitics were not sufficiently accounted for. SNDR (ENOB) is ≥ 39 dB (6.2-b) and more or less constant up to about 4 MS/s. Peak SNDR when f_{in} = 20.13 MHz is 42.0 dB (6.68-b) at 1 MS/s. The spectral performance vs. input magnitude for f_s = 3 MHz and f_{in} = 1.83 MHz is shown in Fig. 10.12. The performance is entirely noise-limited all the way up to 0 dBFS, where clipping occurs due to a large offset. The useful *dynamic range* is therefore 49 dB (7.84-b). Intermodulation distortion at 3 MS/s was measured using a two-tone input signal, f_{in} = {1.73, 1.83} MHz, each at -7.33 dBFS. The third-order spurious were ≤ -48.95 dBc, giving a *third-order intercept* (TOI) equal to +17.1 dBFS. A zero input current sampled at 3 MS/s showed the converters rms *selfnoise* to be 1.24 LSB centered around code 559. This is equal to a noise power of -49.3 dB relative to a full-scale sinusoidal signal. The ADC is fully functional with little or no performance degradation at supply voltages from 2.8 to 3.5 V. A 110 µA bias current was used internally, and the full-scale current was set to +/-50 µA. Analog and digital power dissipation including digital output drivers was measured at 3 MS/s to 82.5 mW from a 3 V supply. The active area is 2.7 mm² when implemented in a 0.8 µm CMOS process.

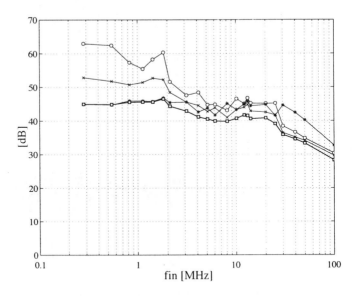

Fig. 10.10 Spectral performance vs. input frequency when f_s = 3 MHz. Legend: (o) SFDR, (x) THD, (*) SNR, and (□) SNDR.

A 3-V wideband CMOS switched-current A/D-converter 143

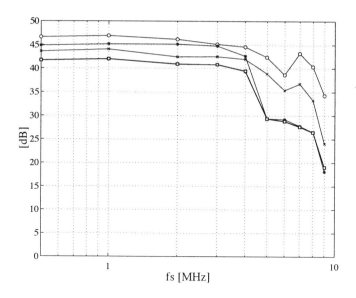

Fig. 10.11 Spectral performance vs. sampling rate when f_{in} = 20.13 MHz. Legend: (o) SFDR, (x) THD, (*) SNR, and (□) SNDR.

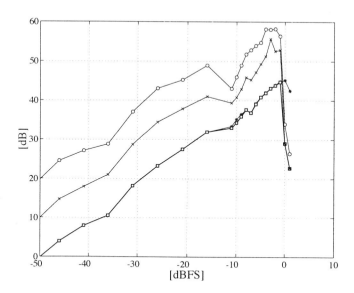

Fig. 10.12 Spectral performance vs. input magnitude with f_s = 3 MHz and f_{in} = 1.83 MHz. Legend: (o) SFDR, (x) THD, (*) SNR, and (□) SNDR.

Parameter	f_{in} = 1.83 MHz, f_s = 3 MHz	f_{in} = 20.13 MHz, f_s = 3 MHz
SFDR	60.3	45.1
THD	-53.8	-42.5
SNR	46.8	44.7
SNDR	46.5	40.8
ENOB	7.43	6.48
	Common parameters	
VDD [V]		3
P [mW]		82.5
A [mm^2]		2.7
RMS Selfnoise		1.24 LSB (-49.3 dBFS)
	Linearity @ f_{in} = 1.13 MHz, f_s = 3 MHz	
DNL @ 10-b		+0.81/–0.9 LSB
INL @ 10-b		+4.9/–4.1 LSB
	Intermodulation @ f_{in} = {1.73, 1.83} MHz	
TOI		+17.1 dBFS

Table 10.2 Summary of experimental results.

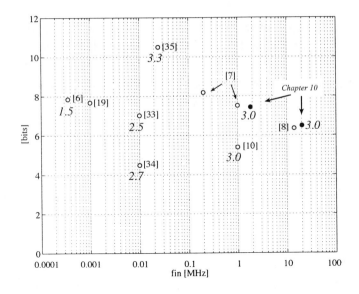

Fig. 10.13 ENOB vs. input frequency for previously reported SI ADCs. Low-voltage designs are indicated with their supply voltages.

10.7 Performance comparison

The measured performance of the presented design is compared with previously reported designs. In signal processing applications it is more important to know the spectral characteristics such as SFDR, SNR and

SNDR rather than the linearity represented by DNL and INL. The following comparisons will therefore focus on the measured effective number-of-bits, calculated from the signal-to-noise-and-distortion ratio as

$$\text{ENOB} = \frac{\text{SNDR} - 1.76}{6.02} \qquad (10.6)$$

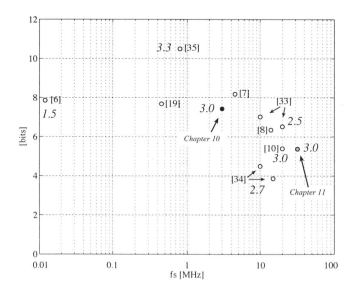

Fig. 10.14 ENOB vs. sampling rate for previously reported SI ADCs. Low-voltage designs are indicated with their supply voltages.

In references where SNDR was not reported, it was estimated from SNR and THD where possible. Since SNR was not given in [19] and [8], ENOB was estimated from THD only. This may be somewhat optimistic. Reference [9] does not report SNR, SNDR or THD measurements, and is therefore omitted from the comparison. An ADC should have a high sampling rate in combination with a large input bandwidth in order to display true wideband performance. Input bandwidth is mandatory while sampling rate can be increased if a parallel ADC architecture is used. The effective resolution vs. input frequency and sample rate for previously reported CMOS SI ADCs is shown in Fig. 10.13 and 10.14. The supply voltage of low-voltage implementations (1.5-3.3 V) is indicated. It is interesting to note that low-voltage designs display both the largest input bandwidth – as in this work – *and* the highest sampling rate [10, 33], while the strongest combination of sampling rate and input bandwidth is achieved by a 5 V design [8]. The design in [34] is a 5-b ADC, which explains the significantly lower ENOB. As expected, there is a decrease in resolution as f_{in} or f_s increases. The highest ENOB (7.51) reported for $f_{in} \geq 1$ MHz is found in [7] and in this work (7.43). The currently highest ENOB (10.5) measured for any SI ADC is reported for the cyclic ADC in [35], using a 24.4 kHz input signal.

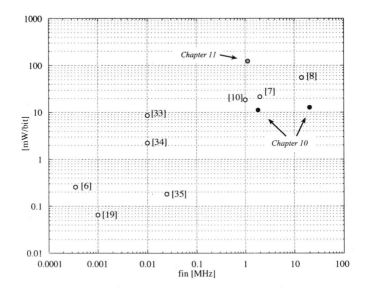

Fig. 10.15 Power dissipation per effective bit resolution vs. input frequency for previously reported SI ADCs.

High bandwidth usually comes with a high power dissipation, and high resolution is often achieved by more complex circuits occupying more chip area. Therefore the area and power efficiency of each design is compared in Fig. 10.15 and 10.16 by plotting the area usage and power dissipation per effective bit resolution vs. the input frequency[2]. The presented ADC is seen to have both a *small area* and *low power dissipation* when compared to other implementations. As a comparison it can also be mentioned that a BiCMOS implementation with ENOB ~ 7.7 for a 10 MHz input sampled at 20 MHz is reported in [36]. It uses a mixed-mode OTA-based S/H. Although the performance of this ADC is superior to that of CMOS implementations, it

---

[2] The comparison could also have been done vs. *sampling* frequency. Many SI ADC designs were reported to have a high sampling rate, while measured performance is only presented for input frequencies that are *much* lower than half the sampling rate, as seen in Fig. 3.4. Therefore *input* frequency was chosen as a more realistic representation of the useful bandwidth of each ADC.

has 18 times the area of this design, and more than 12 times as high power dissipation.

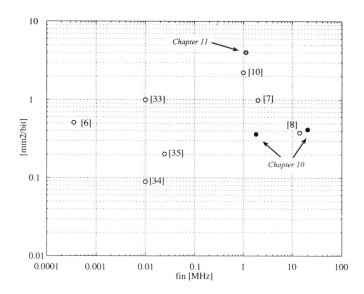

Fig. 10.16 Chip area per effective bit resolution vs. input frequency for previously reported SI ADCs.

10.8 Summary

In this chapter, we have studied a 10-b switched-current A/D-converter design, and made a comparison with previously reported CMOS SI A/D converters. In general, it is seen that low-voltage implementations (2.5 – 3.3 V) are capable of wide input bandwidth as well as a high sample rate. Among wideband ADCs with $f_{in} \geq 1$ MHz, the converter in this chapter has the highest reported SFDR, input bandwidth *and* power efficiency. Its area efficiency and peak ENOB is equivalent to the best of the previously reported implementations while the sample rate is lower. Due to the high input bandwidth, a parallel ADC architecture can be used to achieve a higher sampling rate while maintaining the wideband performance. SNDR is better than 40 dB up to more than 20 MHz, and with a sinusoidal input, $f_{in} = 1.83$ MHz sampled at 3 MHz, the measured SFDR and SNDR are 60.3 and 46.5 dB respectively. The die area is approximately 2.7 mm^2 including digital correction. This 3 V design was implemented in a 0.8 µm, 5 V digital CMOS process demonstrating the low-voltage capability of the switched-current technique. Power dissipation is 82.5 mW including digital circuits and output drivers.

SI circuits are often claimed to have an "inherent high-speed potential" and it has been shown in signal filtering applications that a sample rate of 80 MS/s is possible in practice [5]. This chapter has demonstrated a switched-current A/D converter with high input bandwidth. Nevertheless it is clear that the performance of SI ADCs in CMOS is not yet compatible with that achieved with SC solutions, e.g. [37] where a 10-b, 95 MS/s ADC with ENOB = 8.7 for f_{in} = 20 MHz is reported. The strength of the switched-current technique lies in its compatibility with digital CMOS processes and its low-voltage properties. Although SI ADCs may not find their way into high-end applications such as radio base stations, they are useful for low-cost, one-chip systems with moderate performance requirements.

REFERENCES

[1] C. Toumazou, J. B. Hughes, N. C. Battersby, (Eds.) *SWITCHED-CURRENTS an analogue technique for digital technology*, IEE Circuits and Systems series 5, 1993.

[2] D. G. Nairn, and C. A. T. Salama, "Ratio-Independent Current-Mode Algorithmic Analog-To-Digital Converters", *Proceedings of ISCAS 89,* Portland, Oregon, pp. 250-253, May 1989, IEEE.

[3] J. Robert, P. Deval, and G. Wegmann, "Novel CMOS Pipelined A/D Convertor Architecture Using Current Mirrors", *Electron. Lett.*, Vol. 25, No. 11, pp. 691-692, May 1989.

[4] P. Deval, J. Robert, and M. J. Declercq, "A 14-bit CMOS A/D Converter Based on Dynamic Current Memories", *Proc. of Custom Int. Circ. Conf*, San Diego, California, pp. 24.2/1-4, May 1991.

[5] J. B. Hughes, and K. W. Moulding, "An 8MHz, 80Ms/s Switched-Current Filter", *Proc. of IEEE Int. Solid-State Circ. Conf.*, San Francisco, California, pp. 60-61, Feb. 1994, IEEE.

[6] C.-C. Cheng, and C.-Y. Wu, "Design Techniques for 1.5-V Low-Power CMOS Current-Mode Cyclic Analog-to-Digital Converters", *IEEE Trans. Circuits Syst.–II*, Vol. 45, No. 1, pp. 28-40, Jan. 1998.

[7] C.-Y. Wu, C.-C. Chen, and J.-J. Cho, "A CMOS Transistor-Only 8-b 4.5Ms/s Pipelined Analog-to-Digital Converter using Fully-Differential Current-Mode Circuit Techniques", *IEEE J. Solid-State Circuits*, Vol. 30, No. 5, pp. 522-532, May. 1995.

[8] M. Bracey, W. Redman-White, J. Richardson, and J. B. Hughes, "A Full Nyquist 15 MS/s 8-bit Differential Switched-Current A/D Converter", *Proceedings of ESSCIRC 95*, Lille, France, pp. 146-149, Sept. 1995.

[9] M. Bracey, W. Redman-White, J. B. Hughes, and J. Richardson, "A 70 MS/s 8-bit Differential Switched-Current CMOS A/D Converter Using Parallel Interleaved Pipelines", *Proceedings of 1995 IEEE Region 10 International Conference on Microelectronics and VLSI*, Hong Kong, pp. 143-146, Nov. 1995.

[10]   Y. Sugimoto, and T. Iida, "A Low-Voltage, High-Speed and Low-Power Full Current-Mode Video-rate CMOS A/D Converter", *Proceedings of ESSCIRC 97*, Southampton, UK, pp. 392-395, Sept. 1997.
[11]   N. Tan, *Switched-Current Design and Implementation of Oversampling A/D Converters*, Kluwer, 1997.
[12]   D. G. Nairn, and C. A. T. Salama, "Algorithmic Analog/Digital Convertor Based on Current Mirrors", *Electron. Lett.*, Vol. 24, No. 8, pp. 471-472, Apr. 1988.
[13]   D. G. Nairn, and C. A. T. Salama, "High-Resolution, Current-Mode A/D Convertors Using Active Current Mirrors", *Electron. Lett.*, Vol. 24, No. 21, pp. 1331-1332, Oct. 1988.
[14]   S. J. Daubert, D. Vallancourt, and Y. P. Tsividis, "Current Copier Cells", *Electron. Lett.*, Vol. 24, No. 25, pp. 1560-1562, Dec. 1988.
[15]   D. G. Nairn, and C. A. T. Salama, "A Ratio-Independent Algorithmic Analog-to-Digital Converter Combining Current Mode and Dynamic Techniques", *IEEE Trans. Circuits Syst.*, Vol. 37, No. 3, pp. 319-325, Mar. 1990.
[16]   D. G. Nairn, and C. A. T. Salama, "Algorithmic analogue-to-digital convertors using current-mode techniques", *IEE Proceedings*, Vol. 137, *Pt G*, No. 2, pp. 163-168, Apr. 1990.
[17]   P. Deval, G. Wegmann, and J. Robert, "CMOS Pipelined A/D Convertor Using Current Divider", *Electron. Lett.*, Vol. 25, No. 20, pp. 1341-1342, Sept. 1989.
[18]   K. C. Wong, and K. S. Chao, "A Current-Mode Successive-Approximation Analog-to-Digital Conversion Technique", *Proceedings of IEEE Midwest Symp. Circ. Syst.*, Calgary, Alta., Canada, Vol. 2, pp. 754-757, Aug. 1990, IEEE.
[19]   S.-W. Kim, and S.-W. Kim, "Current-Mode Cyclic ADC for Low-Power and High-Speed Applications", *Electron. Lett.*, Vol. 27, No. 10, pp. 818-820, May 1991.
[20]   D. Macq, and P. G. A. Jespers, "A 10-bit Pipelined Switched-Current A/D Converter", *IEEE J. Solid-State Circuits*, Vol. 29, No. 8, pp. 967-971, Aug. 1994.
[21]   B. E. Jonsson, "A 3 V, 10 bit, 6.4 MHz Switched-Current CMOS A/D Converter Design", *Proc. of Int. Conf. on Electronics Circuits and Systems (ICECS)*, Lisbon, Portugal, Vol. 1, pp. 27-30, Sept. 1998, IEEE.
[22]   C.-C. Chen, C.-Y. Wu, and J.-J. Cho, "A 1.5 V CMOS Current-Mode Cyclic Analog-to-Digital Converter with Digital Error Correction", *Proc. of Int. Symp. Circuits And Systems (ISCAS)*, Seattle, Washington, pp. 537-540, May 1995, IEEE.
[23]   M. Bracey, W. Redman-White, J. Richardson, and J. B. Hughes, "A Full Nyquist 15 MS/s 8-b Differential Switched-Current A/D Converter", *IEEE J. Solid-State Circuits.*, Vol. 31, No. 7, pp. 945-951, July 1996.
[24]   B. Ginetti, P. G. A. Jespers, and A. Vandemeulebroecke, "A CMOS 13-b Cyclic RSD A/D Converter", *IEEE J. Solid State Circ.*, Vol. 27, No. 7, pp. 957-964, July 1994.
[25]   B. Jonsson, and N. Tan, "Clock-Feedthrough Compensated First-Generation SI Circuits and Systems", *Analog Integrated Circuits and Signal Processing*, Vol. 12, No. 4, pp. 201-210, Apr. 1997.
[26]   N. K. Verghese, T. J. Schmerbeck, and D. J. Allstot, *Simulation Techniques and Solutions for Mixed-Signal Coupling in Integrated Circuits*, Kluwer, 1995.
[27]   T. S. Fiez, D. J. Allstot, G. Liang, and P. Lao, "Signal-Dependent Clock-Feedthrough Cancellation in Switched-Current circuits", *Proc. of China 1991 Int. Conf. Circuits And Systems,* Shenzhen, China, pp. 785-788, June 1991, IEEE.
[28]   N. Tan, and S. Eriksson, "Low-Voltage Fully Differential Class-AB SI Circuits with Common-Mode Feedforward", *Electron. Lett.*, Vol. 30, No. 25, pp. 2090-2091, Dec. 1994.

[29] B. E. Jonsson, "Design of Power Supply Lines in High-Performance SI and Current-Mode Circuits", *Proc. of 15th NORCHIP Conf.*, Tallinn, Estonia, pp. 245-250, Nov. 1997, IEEE.

[30] H. Träff, "A Novel Approach to High Speed CMOS Current Comparators", *Electron. Lett.*, Vol. 28, No. 3, pp. 310-312, Jan. 1992.

[31] J. Doernberg, H.-S. Lee, and D. A. Hodges, "Full-Speed Testing of A/D Converters", *IEEE J. Solid State Circ.*, Vol. SC-19, No. 6, pp. 820-827, Dec. 1984.

[32] M. v.d. Bossche, J. Schoukens, and J. Renneboog, "Dynamic Testing and Diagnosis of A/D Converters", *IEEE Trans. Circ. Syst.*, Vol. CAS-33, No. 8, pp. 775-785, Aug. 1986.

[33] M. Gustavsson, *Analog Interfaces in a Digital CMOS Process*, Licentiate Thesis No. 662, Linköping University, Sweden, Dec. 1997.

[34] M. Gustavsson, and N. Tan, "New Current-Mode Pipeline A/D Converter Architectures", *Proceedings of ISCAS 97*, Hong Kong, Vol. 1, pp. 417-420, June 1997, IEEE.

[35] J.-S. Wang, and C.-L. Wey, "A 12-bit 100ns/bit 1.9-mW CMOS Switched-Current Cyclic A/D Converter", *IEEE Trans. on CAS-II*, Vol. 46, No. 5, pp. 507-516, May 1999.

[36] D. Robertson, P. Real, and C. Mangelsdorf, "A Wideband 10-bit, 20Msps Pipelined ADC using Current-Mode Signals", *Proceedings of IEEE Solid-State Circ. Conf.*, San Francisco, California, pp. 206-207, Feb. 1990, IEEE.

[37] K. Y. Kim, N. Kusayanagi, and A. A. Abidi, "A 10-b, 100-MS/s CMOS A/D Converter", *IEEE J. Solid State Circ.*, Vol. 32, No. 3, pp. 820-827, Mar. 1997.

Chapter 11

A Dual 3-V 32-MS/s CMOS Switched-Current ADC[1]

The potential for switched-current A/D converters in low-voltage, telecommunication applications with a high level of integration is investigated through the test design described in this chapter. A dual, 3 V, 32 MS/s A/D converter was fabricated in a standard digital 5 V, 0.8 mm CMOS process. Fully differential first-generation switched-current circuits with common-mode feedforward are used to implement a 1.5-b/stage pipelined ADC core. Eight time-interleaved ADC cores operating at 4 MS/s are used to achieve a high sample rate. With channel compensation, the measured SFDR is more than 50 dB at 32 MS/s with f_{in} = 1.13 MHz. The ADC-core was measured to have 60.3 dB peak SFDR, 46.5 dB peak SNDR, and approximately 20 MHz input bandwidth. The resolution of the parallel ADC was limited by additional noise and the useful bandwidth was lowered by a fixed-pattern timing error that could not be removed by channel calibration.

11.1 Introduction

High-speed, wideband A/D-converters are essential building blocks in wideband radio applications. The purpose of the A/D converter is to digitize the incoming signal as early as possible in the signal processing chain. An OFDM receiver system example is shown in Fig. 11.1. The dual (I and Q) A/D converter is highlighted. The *switched-current* (SI) technique is often

---

[1] B. E. Jonsson, and H. Tenhunen, "A Dual 3-V 32-MS/s CMOS Switched-Current ADC for Telecommunication Applications", *Proc. of 1999 Int. Symp. Circuits and Systems*, Orlando, Florida, Vol. 2, pp. 343-346, May 1999, IEEE.

claimed to have high-speed potential, and sample rates up to 80 MS/s have been demonstrated in signal filtering [1]. Considering the compatibility with digital CMOS technology [2], and the low-voltage capability, the SI technique should be an attractive solution for embedded ADCs in low-cost, wideband applications. Nevertheless, only a few SI implementations with input and sampling frequencies above 1 MHz have been reported [3-7]. A BiCMOS technology was used for the design in [3]. The highest measured sampling rate for previously reported SI ADCs is 20 MHz [3, 6-7]. In this chapter we will examine a 3 V, 32 MS/s ADC implemented in a 5 V digital CMOS process.

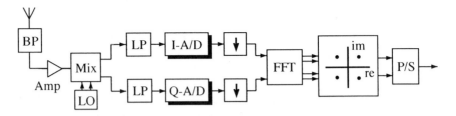

Fig. 11.1 OFDM system example.

11.2   A/D converter architecture

The parallel A/D converter architecture with calibration is shown in Fig. 11.2. Parallel processing is a well-known method to increase throughput that has also been used for switched-current ADCs [6]. The drawback with parallel *analog* processing elements is that complete matching is never achieved in practice. Channel offset will generate a fixed-pattern noise appearing as spurious tones at multiples of f_s/N, where N is the number of channels [6]. Channel gain mismatch results in a fixed-pattern modulation that creates mirror images of the input signal centered around multiples of f_s/N. It is also critical that the sampling time instances are equally spaced since any fixed-pattern timing errors will cause mirror images similar to those created by gain mismatch. Channel gain and offset errors can be removed with a simple calibration scheme, while timing errors are not easily compensated for. *On-chip digital background calibration* was demonstrated in [8]. In this chapter, a simplified, experimental approach is used. A low-frequency sinusoidal input is used to estimate the offset and gain of each channel as

$$os_i = average(d_i)$$
$$G_i = rms(d_i)$$
(11.1)

The estimated parameters are then used to correct the reconstructed signal at all frequencies according to Eq. 11.2.

$$G'_i = G_i/\max(G_i)$$
$$d'_i = (d_i - os_i)/G'_i \qquad (11.2)$$

In order to properly emulate a hardware implementation, the compensation is implemented in MATLAB as

$$d'_i = d_i - os_i + \gamma_i(d_i - os_i)$$
$$\gamma_i = (1 - G'_i)/G'_i \qquad (11.3)$$

where each term is truncated to the resolution of the ADC.

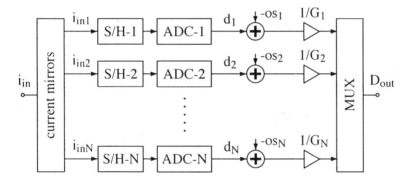

Fig. 11.2 Parallel pipelined ADC with channel gain and offset compensation.

11.3 Circuit implementation

A 10-b, 32 MS/s ADC was implemented in a 0.8 μm, 5 V standard digital process. The die area is approximately 22 mm² including digital circuits. Eight ADCs are operated in parallel, and each sub-ADC is one instance of the 1.5-b/stage pipelined ADC presented in [9] (see also chapter 10). The analog part is implemented with clock-feedthrough (CFT) compensated first-generation SI circuits [10] and current mirrors. A fully-differential approach with common-mode feedforward [11] was chosen in order to suppress external noise as well as signal-independent CFT. A time-interleaved approach inevitably leads to increased layout complexity. An *interdigitized layout*, rather than common-centroid was used to ensure good device matching and a dense layout. Power supply lines were sized according to [12] in order to reduce the local and global drain current mismatch due to resistive voltage drops. Special attention was given the input current mirror generating eight copies of the input current. The current comparator proposed by Träff [13] performs the bit-level decision of each bitcell, and a 1.5-b sub-DAC was realized with current sources and current-steering switches. The digital part consists of on-chip digital correction circuits and a control unit. Due to the complexity of the design, clock distribution and data buses occupy a relatively large portion of the chip. This led to clock timing problems that could be seen in the measurements. A plot of the final layout is shown in Fig. 11.3.

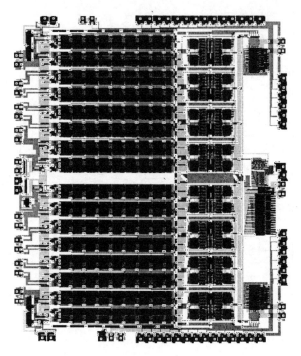

Fig. 11.3 Chip layout of the dual ADC.

11.4 Experimental results

11.4.1 ADC core cell

The performance of each ADC core is analyzed by keeping every eighth output sample. The spectral performance of the ADC core vs. input frequency when sampled at 3 MS/s is shown in Fig. 11.4. Measured SNDR is ≥ 40 dB for input frequencies up to more than 20 MHz. Peak SNDR is

46.5 dB (7.43 bits). The spectral performance vs. sampling rate when f_{in} = 20 MHz is shown in Fig. 11.5. It is seen that the maximum sample rate is approximately 4 MS/s, and therefore a 32 MS/s total sampling rate and an input bandwidth of about 20 MHz is predicted for the parallel structure.

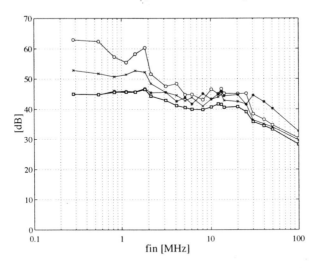

Fig. 11.4 Spectral performance of the ADC core vs. input frequency when f_s = 3 MHz. Legend: (o) SFDR, (x) THD, (*) SNR, and (□) SNDR.

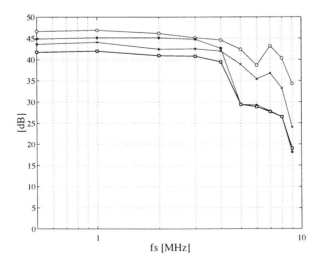

Fig. 11.5 Spectral performance of the ADC core vs. sampling rate when f_{in} = 20.13 MHz. Legend: (o) SFDR, (x) THD, (*) SNR, and (□) SNDR.

11.4.2 Parallel ADC

The ADC output spectrum of a 1.13 MHz input current sampled at 32 MS/s before channel calibration is shown in Fig. 11.6. Channel offset produced the tones seen at 4, 8, 12, and 16 MHz while gain mismatch is responsible for the input images at +/- 1.13 MHz from those frequencies. The largest offset tone is –30 dBc (16 MHz) and the largest gain error tone is -34.7 dBc (2.87 MHz). The effect of channel calibration is shown in Fig. 11.7 where SFDR has increased to more than 50 dB. SNDR is limited by noise to 38 dB, which is equal to 6.0 effective bits. The incomplete cancellation of channel mismatch is partially due to the difference between the low-frequency calibration signal and the actual signal, but mostly caused by a fixed-pattern timing error due to clock skew. Channel timing errors are neither detected nor compensated for by the calibration algorithm since they represent phase errors rather than amplitude errors. The effect of timing errors increases with the increasing slope (or frequency) of the input signal. We can estimate the magnitude of the timing error from the measured data as follows: If the input current, i_{in} is

$$i_{in} = A\sin(2\pi f_{in} t) \tag{11.4}$$

and the timing error power in dBc is

$$\varepsilon_{dBc} = 20\log\left(\frac{rms(\Delta i_{in})}{rms(i_{in})}\right) \tag{11.5}$$

then the rms timing error, σ_t, is estimated from

$$\sigma_t = \frac{10^{\frac{\varepsilon_{dBc}}{20}}}{2\pi f_{in}} \tag{11.6}$$

From Fig. 10.7 we identify timing errors as spurious tones at {–51, –58, –53, –54, –60, –58} dBc, giving the total timing error power $\varepsilon_{dBc} = -46.75$ dBc, which according to Eq. 11.6 is equivalent to a timing error $\sigma_t \approx 0.65$ ns. With a 20 MHz input signal, the SFDR is entirely dominated by timing errors, and the calibration algorithm only removes channel offset effectively. Although the sub-ADCs were measured to have an input bandwidth of more than 20 MHz, the useful bandwidth of the parallel ADC is much less, due to the timing errors. With a redesign of the clock distribution circuits, a high

A dual 3-V 32-MS/s CMOS switched-current ADC

sample rate *and* bandwidth can be expected. Total power dissipation is 2 x 660 mW from a 3 V supply, and the chip area is 2 x 22 mm^2.

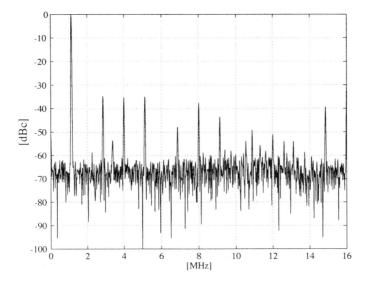

Fig. 11.6 Measured parallel ADC output spectrum before channel calibration.

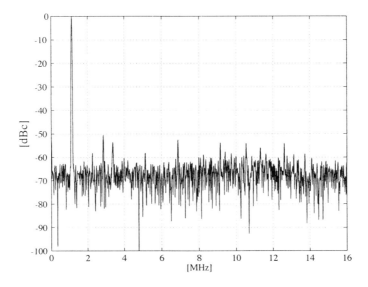

Fig. 11.7 Measured parallel ADC output spectrum after channel calibration.

Parameter	without channel calibration	with channel calibration
SFDR	30	50
SNR	38	38
SNDR	27.7	38
ENOB	4.3	6.0
	Common parameters	
VDD [V]	3	
P [mW]	2 x 660	
A [mm^2]	2 x 22	
f_{in} [MHz]	1.13	
f_s [MHz]	32	
σ_t [ns]	0.65	

Table 11.1 Summary of experimental results.

11.5 Summary

In this chapter we have seen a dual 3 V, 32 MS/s MOST-only ADC implemented in a 0.8 μm, 5 V standard digital process. The die area is approximately 22 mm^2/ADC including digital correction. Power dissipation at 3 V is estimated to 660 mW/ADC, including output drivers. With a sinusoidal input, f_{in} = 1.13 MHz, sampled at 32 MHz, the measured SFDR and SNDR are 50 and 38 dB respectively, after channel calibration. The measurements display the highest sampling rate currently reported for any switched-current ADC, and the input bandwidth and resolution of the ADC core are equivalent to that of previously reported implementations [4-5]. Insufficient design of the on-chip clock distribution bus, however, led to a fixed-pattern timing error that reduced the useful input bandwidth of the parallel ADC. From the results presented in this chapter and the measured results of previously reported implementations [4-7] it is clearly seen that CMOS switched-current A/D converters are not yet suitable for wideband radio applications, where a high resolution is required. The strength of the SI technique lies mainly in the compatibility with digital CMOS technology which gives the potential for low-cost, embedded A/D-converters with moderate resolution.

REFERENCES

[1] J. B. Hughes, and K. W. Moulding, "An 8MHz, 80MS/s Switched-Current Filter", *Proceedings of IEEE Solid-State Circ. Conf.,* San Francisco, California, pp. 60-61, Feb. 1994, IEEE.

[2] C. Toumazou, J. B. Hughes, N. C. Battersby (Eds.), *SWITCHED-CURRENTS an analog technique for digital technology,* IEE Circuits and Systems series 5, 1993.

[3] D. Robertson, P. Real, and C. Mangelsdorf, "A Wideband 10-bit, 20Msps Pipelined ADC using Current-Mode Signals", *Proceedings of IEEE Solid-State Circ. Conf.,* San Francisco, California, pp. 206-207, Feb. 1990, IEEE.

[4] C.-Y. Wu, C.-C. Chen, and J.-J. Cho, "A CMOS Transistor-Only 8-b 4.5Ms/s Pipelined Analog-to-Digital Converter using Fully-Differential Current-Mode Circuit Techniques", *IEEE J. Solid-State Circuits.,* Vol. 30, No. 5, pp. 522-532, May. 1995.

[5] M. Bracey, W. Redman-White, J. Richardson, and J. B. Hughes, "A Full Nyquist 15 MS/s 8-bit Differential Switched-Current A/D Converter", *Proceedings of ESSCIRC 95,* pp. 146-149, 1995.

[6] M. Bracey, W. Redman-White, J. B. Hughes, and J. Richardson, "A 70 MS/s 8-bit Differential Switched-Current CMOS A/D Converter Using Parallel Interleaved Pipelines", *Proceedings of 1995 IEEE Region 10 Int. Conf. on Microelectronics and VLSI,* Hong Kong, pp. 143-146, 1995.

[7] Y. Sugimoto, and T. Iida, "A Low-Voltage, High-Speed and Low-Power Full Current-Mode Video-rate CMOS A/D Converter", *Proceedings of ESSCIRC 97,* Southampton, UK, pp. 392-395, Sept. 1997.

[8] D. Fu, K. Dyer, S. Lewis, and P. Hurst, "Digital Background Calibration of a 10b 40Msample/s Parallel Pipelined ADC", *Proceedings of IEEE Solid-State Circ. Conf.,* San Francisco, California, pp. 140-141, Feb. 1998, IEEE.

[9] B. E. Jonsson, and H. Tenhunen, "A 3V Switched-Current Pipelined Analog-to-Digital Converter in a 5V CMOS process", *Proc. of 1999 Int. Symp. Circuits and Systems,* Orlando, Florida, Vol. 2, pp. 351-354, May 1999, IEEE.

[10] T. S. Fiez, D. J. Allstot, G. Liang, and P. Lao, "Signal-Dependent Clock-Feedthrough Cancellation in Switched-Current circuits", *Proc. of China 1991 Int. Conf. Circuits And Systems,* Shenzhen, China, pp. 785-788, June 1991, IEEE.

[11] N. Tan, and S. Eriksson, "Low-Voltage Fully Differential Class-AB SI Circuits with Common-Mode Feedforward", *Electron. Lett.,* Vol. 30, No. 25, pp. 2090-2091, Dec. 1994.

[12] B. E. Jonsson, "Design of Power Supply Lines in High-Performance SI and Current-Mode Circuits", *Proc. of 15th NORCHIP Conf.,* Tallinn, Estonia, pp. 245-250, Nov. 1997, IEEE.

[13] H. Träff, "A Novel Approach to High Speed CMOS Current Comparators", *Electron. Lett.,* Vol. 28, No. 3, pp. 310-312, Jan. 1992.

Chapter 12

Conclusions

It has been about twelve years since the advent of the switched-current technique. In the early days *anything* seemed possible: high speed, low voltage, inherently mismatch-free building blocks, and even low power dissipation were the predicted benefits of this new and exciting technique. Because no other components than MOS transistors were required and the dynamic range would no longer be hard-limited by the supply voltage, a technique that was truly compatible with the rapid development in digital VLSI technology had been found. With this technique it became possible to integrate analog interfacing circuits, A/D and D/A, with the digital signal processing circuits in order to have cost effective one-chip solutions. As the SI technique started its maturing process, it became evident that it also had a number of problems to be solved: Clock-feedthrough (CFT) or charge-injection from sampling switches was far more dominant than it had ever been in SC circuits. Thermal noise also seemed to be much more present in current-mode circuits. Would it ever be possible to achieve a signal-to-noise ratio equivalent to that in voltage-mode circuits? Accurate matching of drain currents became more important than before, and also the linearity of a current mirror. Another significant issue was to develop new realizations of necessary building blocks with *current* input and output.

A lot of work has been done by a large number of research groups in order to solve these problems. One of the main areas has been the cancellation or compensation of clock-feedthrough. Many different approaches have been proposed, and among them the first scheme that could theoretically achieve *complete* cancellation of CFT was proposed by the author. Another novel circuit realization proposed by the author is the current-mode N-port adaptor for SI wave filters. Several publications have shown the tradeoffs necessary in designing a low-noise, low-distortion SI circuit. First, clock-feedthrough and noise must be carefully analyzed. For high-performance SI circuits it is also essential to pay close attention to small details. The author's contribution to this is captured in the two chapters on sampling jitter and resistive voltage drop on power supply lines. Ultimately, the usefulness of any new technique must be tested through physical implementation. A large number of circuit implementations have been reported in the open literature, particularly within the fields of signal filtering and data conversion. Early work by the author includes a $\Delta\text{-}\Sigma$ modulator implementation using the first generation SI memory cells proposed by the author. Its measured 11-b performance was unusually high at the time of publication, and demonstrated the quality of the memory cell. A wideband ADC suitable for undersampling or parallel time-interleaved applications was also implemented by the author. It was measured to have a 20 MHz input bandwidth and a peak resolution of 7.43 effective bits at 3 MS/s. Although this is the highest bandwidth reported for any CMOS SI ADC, and the second highest resolution for implementations with $f_{in} \geq 1$ MHz, it is still not compatible with what has been achieved with switched-capacitor solutions. By implementing eight of these ADCs in parallel, a sampling rate of 32 MHz was achieved. While being the highest sample rate reported for any SI ADC in the literature, it is still significantly less than what has been reported for switched-capacitor solutions.

It seems as if the first enthusiasm regarding the switched-current technique has slowly dwindled before having any serious influence on the commercial microelectronics area. There are good reasons for that: First and foremost, SI circuits have not yet been able to display the same linearity and noise performance as their voltage-mode counterparts. Low voltage, or at least reasonably low voltage, *has* been demonstrated, but that has been done with voltage-mode circuits as well. When it comes to mixed-signal integration, the potential is still there but has not been fully demonstrated. This has nothing to do with the SI technique itself but is more due to the difference between analog and digital design methodology and the lack of good mixed analog-digital CAD tools. Basically, digital design of today is a different universe. Designers use VHDL code, top-down synthesis, and in many cases the actual layout is not seen by the designer since layout synthesis is often done by the silicon supplier. Analog design, on the other hand, is still predominantly a bottom-up activity. The physical layout is such an essential factor in reaching the final performance that it is usually done by the design group. Most often, the analog and digital designers are not the same persons, and their optimum choices of CAD tools are very different. Today, digital design is essentially a programming activity and, while the digital designer needs a good debugging tool, the analog designer must have

Conclusions

an accurate circuit-level simulator with detailed analog models of the CMOS process. Often in practice, these worlds just do not meet. This is one of the reasons why the cornerstone in the marketing of SI circuits, the compatibility with constantly scaled state-of-the-art digital CMOS has not been fully demonstrated – at least not enough to convince a majority of microelectronics companies of its benefits.

The regularity of SI circuits is still unexplored. There is no comparison to the simplicity with which SI circuits can be automatically generated. Surprisingly few publications deal with this issue in spite of its obvious advantages. In this sense, SI circuits have more in common with digital gates than with analog circuits such as OP-amps. A reasonably small effort in CAD tool development would enable the synthesis of SI interfacing circuits much in the same way as digital circuits are synthesized. This is yet to be done. With such CAD tools available, and after a few more generations of CMOS scaling, the SI technique is likely to emerge once again as an alternative for low-cost, medium performance, embedded analog interfacing circuits.

In order to truly find out the limits and potentials of the SI technique, it should be put to test in commercial projects. It is a well-known fact from voltage-mode A/D-converters that commercially available implementations usually outperform the academic research by at least a factor of two. In industry, more resources can be spent on refining each circuit, and the further development of the SI technique would benefit from that. Some work remains to be done in order to develop an attractive design methodology for SI circuits. Particularly, the potential for design automation should be fully explored. Automated synthesis and layout generation tools may be what it takes for SI circuits to penetrate the microelectronics industry a few years into the 21st century.

Appendix

Noise Integrals

These integrals may be helpful when determining the total white-noise power transmitted through a linear system with the frequency response $H(f)$.

$H(f)$	$\int_0^\infty |H(f)|^2 df$
$\dfrac{1}{1+\dfrac{f}{f_0}}$	$f_0 \dfrac{\pi}{2} = \dfrac{\omega_0}{4}$
$\dfrac{1}{\left(1+\dfrac{f}{f_0}\right)\left(1+\dfrac{f}{f_1}\right)}$	$\dfrac{f_0 f_1}{f_0 + f_1} \cdot \dfrac{\pi}{2} = \dfrac{\omega_0 \omega_1}{\omega_0 + \omega_1} \cdot \dfrac{1}{4}$
$\dfrac{\left(1+\dfrac{f}{f_{z0}}\right)}{\left(1+\dfrac{f}{f_0}\right)\left(1+\dfrac{f}{f_1}\right)}$	$\left(1+\dfrac{f_0 f_1}{(f_{z0})^2}\right)\dfrac{f_0 f_1}{f_0 + f_1}\cdot\dfrac{\pi}{2} = \left(1+\dfrac{\omega_0 \omega_1}{(\omega_{z0})^2}\right)\dfrac{\omega_0 \omega_1}{\omega_0 + \omega_1}\cdot\dfrac{1}{4}$

Index

Index

A/D
    1.5-b pipeline 132
    algorithmic 131
    circuit-level simulation 138
    conversion 55
    converter 1, 12, 13, 18, 49, 54, 55, 91, 94, 130, 131, 146, 151
    current-division 59
    cyclic 56
    dual I/Q 154
    flash 58
    floorplanning 99
    folding 59
    interstage gain 57
    layout 102, 135
    oversampling 59, 121
    parallel 57, 152, 156
    pipeline 56
    successive approximation 59
    ultra-low voltage 58
    wideband 57, 129, 151
active mirror 18
amplifier
    common-gate 34
    current 3
    interstage 132
    OTA 12

cascode 18, 112, 115
    branch 115
    layout 100
    low-voltage 18, 80, 100
channel
    calibration 58, 156
    channel length modulation 10, 16
    charge 20, 34
    charge redistribution 19, 22
    effective length 19
    gain mismatch 152
    gate-channel capacitance 19, 20
    mismatch 58, 156
    offset 57, 152, 156
    switch 21
    timing errors 156
class-AB 39, 50, 51
clock 10, 11
clock bus 98, 100
clock distribution 154, 156
clock edge 20, 22, 34, 85
clock reference 85
clock signal coupling 19
clock skew 156
clock timing 34, 154
clock-feedthrough 3, 9, 19, 22, 32, 71, 113, 124, 133, 137
    compensation 32, 72
    complete cancellation 74
    equivalent gain error 74
    error current 74
    large signal analysis 73
    model 22, 23
    parasitics 19
    voltage 74
CMOS switch 34
common-centroid 15, 133, 153
common-mode
    error 37
    feedback 37
    feedforward 37, 129, 133, 134, 153
    rejection 37, 38
comparator
    current comparator 12, 13, 122, 124, 133, 154
    offset 132, 137
conductance
    output 36, 133
    ratio 3, 18, 32, 80, 124, 133, 137
    switch 30, 31

D/A
    conversion 9
    converter 1, 13, 91
    converter 1.5-b 134
    converter 1-b 14, 59, 122, 124
delta-sigma 54, 59, 121
distortion
    distortion-limited 138
    harmonic 14, 32, 34, 74, 80, 84, 85, 91, 125, 140
    intermodulation 142
    low-distortion 91
    non-linear 18
    SFDR 137, 138, 156
    SNDR 125, 126, 138, 140, 154, 156
    THD 51, 72, 140
dummy circuit 34
dummy switch 33, 36, 38

Index

dynamic range 2, 60, 86, 125, 126, 142

effective number-of-bits 145

filter 9, 14, 18, 49, 109
    adaptor 112
    automatic design 97
    bilinear 51
    biquad 51
    coefficients 117
    design & analysis 53
    elliptic 117
    FIR 49
    IIR 50
    integrator based 51
    ladder 51
    median 53
    multirate 53
    programmable 51
    reference filter 116
    wave filter 109, 110
    wave filters 52
fully differential 37, 38, 83, 84, 87, 89, 129, 133, 153

histogram 123

impedance
    characteristic 111
    input 18
    input (comparator) 13
    output 18
interdigitized 15, 133, 153
intermodulation 61, 142

layout
    guard bar 100
    style 100

mismatch
    bias current 32
    current 91
    local 91
    threshold voltage 17
    transistor 15, 72, 137

noise
    calculations 25
    circuit noise 24, 26
    current 25, 27
    external 103, 133, 153
    flicker noise 25
    in-band 61
    noise-limited 140, 142, 156
    phase-noise 83
    power 24
    PSD 24
    quantization 59
    rms 26
    sampled 25
    selfnoise 142
    shot noise 25
    SNR 28, 29, 60, 85, 86, 127
    source 25, 27
    substrate 99, 133, 136
    thermal 1, 25, 136
    transfer function 61
    transistor 24
    white 25, 26
    voltage 26

power
    supply 2, 91
    supply separation 99
    supply wire 31, 91, 92, 102, 133, 153
    supply wire width 95

routing channel 101

sampling
    correlated double 25
    fully differential 88
    jitter 31, 32, 83, 84, 85
    jitter reduction 86
    S/H 9, 12, 25, 77, 133, 134, 138, 146
    S/H evolution 39
    signal dependent jitter 87
    switch 1, 32, 38, 83, 98, 130
simulation
    circuit level 30, 53, 136, 138, 140
    HSpice 118
    Matlab 86, 88, 95, 137, 138, 140, 153
    Spectre 118, 138
    system level 123, 136
single ended 89

Index

technology
  BiCMOS 12, 57, 146, 152
  CMOS 2, 14, 80, 120, 125, 129, 130,
    133, 142, 148, 151, 152, 158
  GaAs 51
timing 83
timing error 152, 156

V/I converter 140
voltage
  drop 91
  swing 13, 24, 29, 83, 84, 86, 87, 88
  voltage mode 2, 50, 83, 84

zero-voltage switching 32, 38, 87